Two Manuals Combined into One Book

Military Police Operations

And

Security Force Assistance

Department of Defense

Pentagon
Publishing

Field Manual
No. 3-19.1

Headquarters
Department of the Army
Washington, DC, 22 March 2001

Military Police Operations

Contents

Distribution Restriction: Approved for public release; distribution is unlimited.

*This publication supersedes FM 19-1, 23 May 1988.

Preface

The Military Police (MP) Corps supports the commander across the full spectrum of military operations. This manual is the foundation for all MP doctrine as it relates to this support. It communicates to all levels of leadership and staffs how the MP provide a flexible and lethal force capable of operating across this full spectrum. As the keystone manual, it identifies what the MP train on and how their forces are organized and equipped in support of all Army echelons. Additionally, this manual provides guidance that can be used by United States (US) Army service schools, students, sister services, and federal agencies.

This manual is based on the purpose, organization, responsibilities, and goals of the US Army as set forth in Field Manuals (FMs) 100-1 and 3-0, as well as corps, division, and brigade manuals. Additionally, this manual is fully compatible with current joint, multinational, and interagency doctrine.

Appendix A contains a metric conversion chart.

The proponent of this publication is HQ TRADOC. Send comments and recommendations on Department of the Army (DA) Form 2028 directly to Commandant, US Army Military Police School (USAMPS), ATTN: ATSJ-MP-TD, 401 Engineer Loop, Suite 2060, Fort Leonard Wood, Missouri 65473-8926.

Unless this publication states otherwise, masculine nouns or pronouns do not refer exclusively to men.

Chapter 1

Introduction

The MP Corps has a strong history evolving over the past five decades. We, as a corps, continue to transform our organizations and doctrine as we have in the past to support the Army in the active defense of the 1970s, the AirLand battle of the 1980s, and now the force-projection doctrine of the 1990s. Our five MP functions clearly articulate the diverse role the MP play across the full spectrum of military operations. We cannot bask in our successes, nor reflect or celebrate. Our charter is to continue our legacy of stellar performance and strive to perfect it.

BG Donald J. Ryder

When the Army developed the Active Defense strategy in 1976, the US was facing the Cold War scenario of central Europe. Military strategy and doctrine were related to a single, focused threat that revolved around the countries in the Warsaw Pact. We were an outnumbered and technically inferior force facing an armor-dominated European battlefield. The MP Corps supported the Active Defense strategy by tailoring its forces to meet the threat. In 1982, when the AirLand Battle strategy was developed, US forces were still outnumbered, but were no longer technically inferior. Still threat-based and focused on a central European conflict, the AirLand Battle strategy used a relatively fixed framework suited to the echeloned attack of soviet-style forces. It delineated and clarified the levels of war; emphasized closed, concerted operations of airpower and ground forces; balanced the offense and the defense; and highlighted the synchronization of close, deep, and rear operations. MP doctrine kept pace with the Army's AirLand Battle strategy by supporting the battlefield commander through four basic missions—battlefield circulation and control, area security (AS), enemy prisoner of war (EPW), and law and order (L&O).

OVERVIEW

1-1. In October 1983, MP capabilities in the AirLand Battle strategy were tested during operation Urgent Fury in Grenada. The MP performed missions that ranged from assisting the infantry in building-clearance operations to assisting Caribbean peacekeeping forces in restoring L&O. These actions secured the MP's place in the combat-support (CS) role, demonstrating the professional knowledge and flexibility necessary for rapid transition from combat to CS to peacetime missions. The changing battlefield conditions of operation Urgent Fury set the stage for the demand of MP units today.

1-2. Evolving simultaneously with the changing definition of the modern battlefield, MP performance in Operations Hawkeye, Just Cause, and Desert Shield/Storm galvanized their ability to perform at any point along the operational continuum. With the publication of FM 100-5 in 1993, the Army adopted the doctrine of full-dimensional operations, relying on the art of battle command to apply those principles and to shift the focus from AirLand Battle to force-projection doctrine. This new doctrine was based on recent combat experience in a multipolar world with new technological advances. Already trained and expected to perform in this new strategy, MP support was already in place and fully operational. The MP continued to perform their basic battlefield missions and to refine their capabilities while supporting the battlefield commander as he deployed to contingency operations throughout the world.

1-3. In 1996, the MP Corps went through a doctrinal review process to determine if it was properly articulating its multiple performance capabilities in support of US forces deployed worldwide (see Appendix B). The review process identified the need to restructure and expand the EPW mission to include handling US military prisoners and all dislocated civilians. This new emphasis transformed the EPW mission into the internment and resettlement (I/R) function. The review process also identified the need to shift from missions to functions. In the past, the four battlefield missions adequately described MP capabilities in a mature theater against a predictable, echeloned threat. However, that landscape is no longer valid. Accordingly, the four MP battlefield missions have become the following five MP functions:

- Maneuver and mobility support (MMS).
- AS.
- L&O.
- I/R.
- Police intelligence operations (PIO).

1-4. These new MP functions are shaped by the following factors:

- The application of stability and support operations where the integration of joint, multinational, and interagency capabilities are common occurrence.
- The lack of traditional linear battlefields, requiring theater commanders in chief (CINCs) to request forces that meet a specific function to accomplish operational requirements.
- The impact of asymmetric threats (such as drug traffickers and terrorist factions) and the effects of man-made and natural disasters.
- The impact of advances in information and communication technologies and specifically in understanding the increased vulnerabilities presented by these technologies.

1-5. Articulating MP capabilities along functional lines benefits the MP and the Army echelon commander as well as the combatant commander. Since there is a multinational, interagency, and sister-service overlap of security services, the importance of including MP leaders and staffs early in the operational planning process cannot be overemphasized. This means before units are designated, unit boundaries are drawn, and unit missions are assigned. Early involvement ensures the proper development of common security responsibilities, communication and connectivity, liaisons, processes, and the rules of interaction between all forces. The ultimate goal should be the optimal, phased employment of MP forces in support of a commander's operational plan. MP functions not only reflect and capture current capabilities, they define the MP Corps in the twenty-first century.

1-6. As the Army reshapes and focuses its resources on transformation, Force XXI, and other redesign efforts, the MP Corps stands proud and ready to support this progress and reiterate its commitment to assist, protect, and defend.

OPERATIONAL FRAMEWORK

1-7. The operational framework consists of the arrangement of friendly forces and resources in time, space, and purpose with respect to each other, the enemy, or the situation (see Figure 1-1). The operational framework for Army forces (ARFOR) rests within the combatant commander's theater organization. Each combatant commander has an assigned geographical area of responsibility (AOR), also called a theater, within which he has the authority to plan and conduct operations. Within the theater, joint-force commanders at all levels may establish subordinate operational areas such as areas of operation (AOs), joint-operations areas (JOAs) and joint rear areas

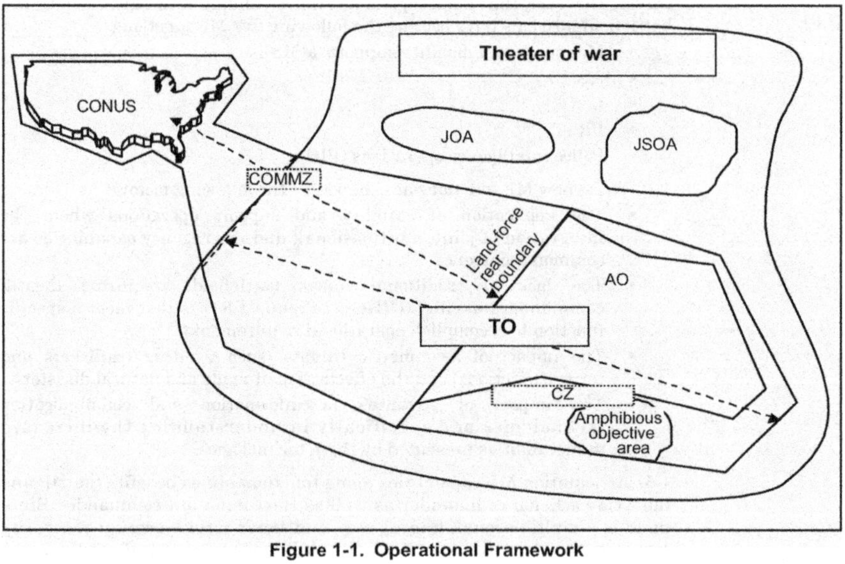

Figure 1-1. Operational Framework

(JRAs). The JRAs facilitate the protection and operation of bases, installations, and forces that support combat operations. When warranted, combatant commanders may designate theaters of war, theaters of operations (TOs), combat zones (CZs), and communications zones (COMMZs).

1-8. A theater of war is that area of air, land, or water that is, or may become, directly involved in the conduct of the war. A theater of war may contain more than one TO. It does not normally encompass the geographic combatant commander's entire AOR. A TO is a subarea (defined by a geographic combatant commander) within a theater of war in which specific combat operations are conducted or supported.

1-9. A CZ is the area required by combat forces for conducting operations. It normally extends forward from the land force's rear boundary. The COMMZ is the rear part of the TO (behind but contiguous to the CZ) that contains the lines of communication (LOC) and provides supply and evacuation support. Other agencies required for the immediate support and maintenance of field forces may also be located in the COMMZ. The COMMZ spans back to the continental US (CONUS) base, to a supporting combatant commander's AOR, or both.

1-10. An AO is an operational area defined by the joint-force commander for land and naval forces. An AO does not typically encompass the entire operational area of the joint-force command (JFC), but it should be large enough for component commanders to accomplish their mission and protect their forces. Army commanders use control measures to describe AOs and to design them to fit the situation and take advantage of the joint force's capabilities. Commanders typically subdivide the assigned AO by assigning subordinate-unit areas. These subordinate-unit areas may be contiguous or noncontiguous (see Figure 1-2). When friendly forces are contiguous, a boundary separates them. When friendly forces are noncontiguous, the concept of operations links the force's elements, but the AOs do not share a boundary. The intervening area between noncontiguous AOs remains the responsibility of the higher headquarters.

BATTLEFIELD ORGANIZATION

1-11. Battlefield organization is the arrangement of forces according to purpose, time, and space to accomplish a mission. Battlefield organization has both a purpose- and spatial-based framework. The purpose-based framework centers on decisive, shaping, and sustaining (DSS) operations. Purpose unifies all elements of the battlefield organization by providing the common focus for all actions. However, forces act in time and space to accomplish a purpose. The spatial-based framework includes close, deep, and rear areas. Despite the increasing nonlinear nature of operations, there may be situations where commanders describe DSS operations in spatial terms. Typically, linear operations involve conventional combat and concentrated maneuver forces. Ground forces share boundaries and orient against a similarly organized enemy force. In such situations, commanders direct and focus simultaneous DSS operations in deep, close, and rear areas, respectively (see FM 3-0).

1-12. The MP battlefield organization supports every Army echelon, from the Army service component command (ASCC) and the theater-support command

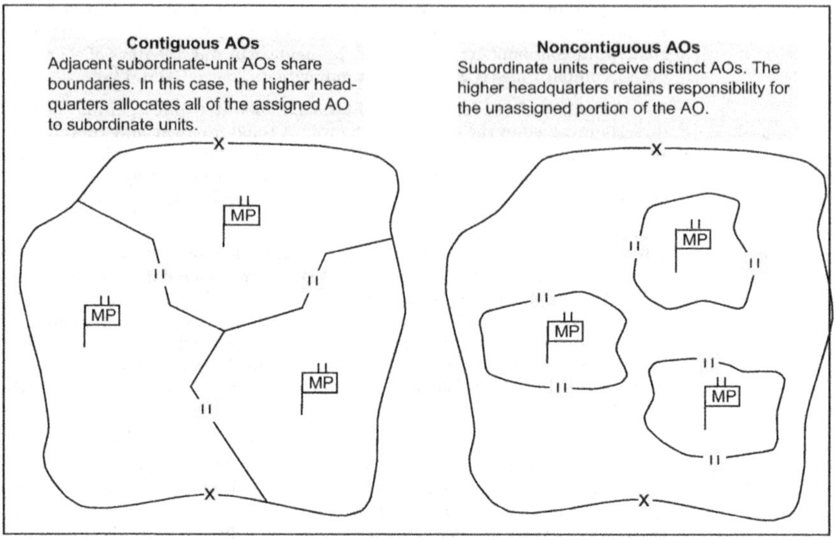

Figure 1-2. Contiguous and Noncontiguous AOs

(TSC) to the maneuver brigade. Regardless of the battlefield organization (purpose or spatial based), MP support to the Army commander is based on available resources and mission, enemy, terrain, troops, time available, and civilian considerations (METT-TC).

1-13. MP support throughout the theater of war may include MP units in the JOA and in the TO. If the combatant commander designates a COMMZ and a CZ within his TO, MP support will come from the established MP modified table of organization and equipment (MTOE) at the subordinate-command echelon. MP support to the JOA is also provided based on METT-TC and available MP assets. Typical MP support may include an I/R brigade liaison detachment (BLD), MP brigades and battalions, a division MP company, a military-working-dog (MWD) team, an L&O team, and a customs team. Figure 1-3, page 1-6, depicts a typical MP organization throughout the TO. In the COMMZ, Figure 1-3 depicts the different types of MP units that are assigned to echelons above corps (EAC) (the ASCC or the TSC). In the CZ, Figure 1-3 depicts the different types of MP units that are assigned to corps, division, and the separate brigades.

SUPPORT IN THE COMMZ

1-14. MP support in the COMMZ is provided by an array of multifunctional MP units. The following MP units provide MP support to EAC:

- The MP brigade (I/R). The MP brigade (I/R) may augment the ASCC or the TSC during wartime. Its mission is to provide command, staff

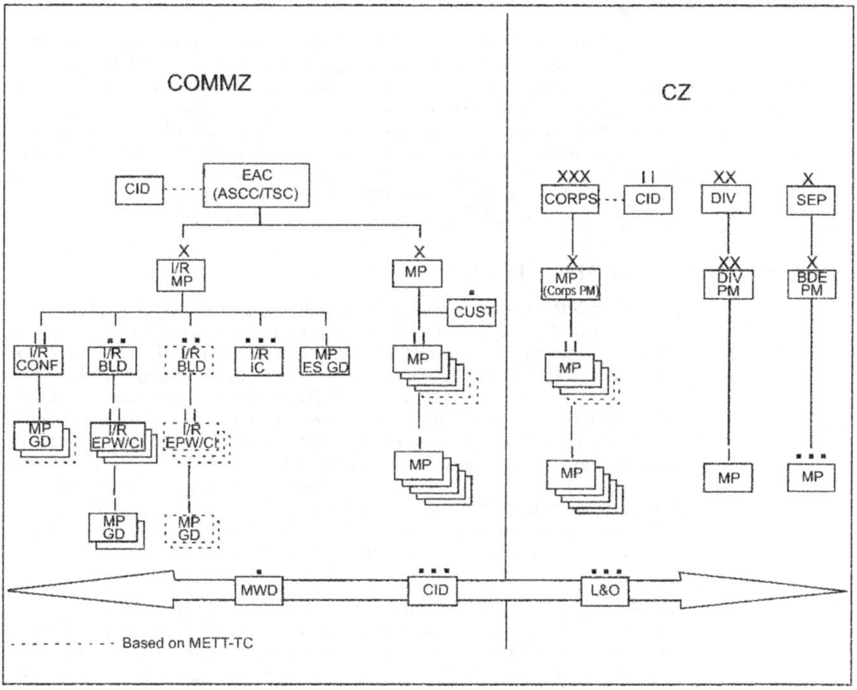

Figure 1-3. MP Structure in the TO

planning, and supervision of I/R operations. This includes coordination with joint and host-nation (HN) agencies, civilian police authority, nongovernmental organizations (NGOs), and US federal agencies.

- The MP brigade (CS). The MP brigade (CS) is assigned to the ASCC or the TSC during wartime (based on METT-TC). The MP brigade (CS) is capable of performing all five MP functions.
- The Criminal Investigation Division (CID) group. The CID group is a stovepipe organization that reports directly to the Commander, US Army Criminal Investigation Command (USACIDC). The CID group provides support to the ASCC and subordinate commands (TSC, corps, or division). See Chapter 9 for further discussion of CID support.

1-15. MP support to other EAC subordinate commands is performed only if MP resources are available. See Chapter 5 for further discussion of MP support to EAC.

SUPPORT IN THE CZ

1-16. MP support is provided in the CZ to each corps, division, and brigade (separate teams or initial/interim brigade combat teams [IBCTs]). An MP brigade (CS) is assigned to each corps, and the MP brigade commander is the corps's provost marshal (PM). A PM and his section, along with an organic division MP company, are assigned to each division. A PM cell and an MP platoon are organic to a separate brigade. A two-person PM cell is organic to the IBCT. The MP units assigned to corps, divisions, and separate brigades are capable of performing all five MP functions. They provide combat, CS, and combat-service-support (CSS) operations within their command's AO.

TYPES OF MILITARY POLICE UNITS

1-17. Most MP units supporting a TO and a JOA are capable of performing all five MP functions. However, the functions must be prioritized based on METT-TC and the availability of MP assets. Current MP structures are designed and tailored to better support the level of command deployed. For example, at the division level, division MP companies are organized as light, heavy, airborne, or air assault and are organic to their respective divisions. The EAC and corps MP brigades and battalions are equally designed to command and control a force mix of up to six battalions or companies. An MP escort-guard and guard company are designed to transport, guard, and provide security to EPWs, civilian internees (CIs), or dislocated civilians. The MP escort-guard company is assigned to the MP brigade (I/R), and the MP guard company is assigned to the MP battalion (I/R).

1-18. MP units can also be tailored and augmented to accomplish multiple, diverse, or specific missions. Customs, L&O, and MWD teams are examples of MP capabilities and flexible responses to a combatant commander's operational requirements. (See Table 1-1, pages 1-9 through 1-19, for a more complete description of MP units. See FM 19-10 and Army Regulation (AR) 190-12 for further information.) The battlefield workload analysis (BWA) is a tool used to determine the number of MP units required to perform some of these multiple missions (see Appendix C).

JOINT, MULTINATIONAL, AND INTERAGENCY OPERATIONS

1-19. In today's environment, the Army will rarely operate or fight alone. The high probability that the Army will operate in concert with its sister services, in an alliance with the forces of foreign nations, or in support of United Nations (UN) operations (when it is committed) is fully reflected in joint doctrine. In such operations, protecting LOC, key facilities, and command and control (C^2) centers will be a shared responsibility. Under this framework, MP units can expect to share the AO with joint, combined, multinational, or interagency resources. MP forces must be prepared to conduct a number of full-spectrum operations with a variety of government and nongovernmental agencies, other services, allied nations, and international agencies.

1-20. Corps and division commanders and staffs must plan (in advance) the transition from a single-service headquarters with joint representation to a joint headquarters capable of functioning as a joint task force (JTF) headquarters. When tasked to form a JTF headquarters, the corps or division must ensure that all of the staff sections and agencies have joint representation (see FMs 100-15 and 71-100). To this end, MP planners must ensure that the JTF is augmented with the appropriate MP forces and with the appropriate echeloned C^2.

1-21. Regardless of the force mix, the MP provide the force with unparalleled, multifunctional capabilities. Among these capabilities is the MP's ability to generate firepower or to handle populations such as EPWs/CIs, dislocated civilians, and refugees. Additionally, MP expertise in investigations and law enforcement enhances the capabilities of other joint, multinational, and interagency police and security forces.

1-22. MP security plans must reflect the joint synergy derived from combining the multiple and diverse capabilities of all participants. To capitalize upon that synergy, MP leaders must keep an open line of communication and coordination to offset the challenges presented by interoperability. Some of these interoperability challenges include—

- Differing political objectives.
- Differing capabilities.
- Cultural/language differences.
- Legal and policy constraints.
- Media impacts.
- Compromise of sensitive processes, procedures, and equipment.
- C^2.
- Communications (digital- and analog-equipment differences).

1-23. MP plans must also accommodate differences in planning capabilities, as well as differences in doctrine, training, and equipment. The intent is to match security missions with force capabilities. MP leaders must understand that operations will often involve multinational teams. While US forces routinely task-organize, this may be more difficult to accomplish with some multinational security forces. This kind of orchestration requires employing standardized procedures, communications, equipment, and liaison within the constraints of operations security (OPSEC).

1-24. Coordination is the key to mission accomplishment in multinational and interagency operations. A military coordination center or a civil-military operations center (CMOC) may meet this coordination requirement. The CMOC provides access for nonmilitary agencies desiring military (to include MP/CID) assistance and coordination. These nonmilitary agencies may include—

- Government organizations (GOs).
- NGOs.
- International organizations (IOs).
- International humanitarian organizations (IHOs).
- HN authorities and agencies.

1-25. The introduction of US Army MP forces in any joint, multinational, or interagency operation is based on METT-TC and the capabilities they bring to the operation. Effective integration of MP forces with other security forces reduces redundant functions, clarifies responsibilities, and conserves resources.

Table 1-1. Unit Descriptions

MP Unit	Mission	Action Elements	Support Role	Capabilities (Include But Are Not Limited to [See TOEs for Details])	Augmentation Needed for Mission Accomplishment
DIVISION MP UNITS					
MP Company (Airborne Division)	Perform the 5 MP functions.	A PM cell and 4 platoons. Each platoon has 3 squads and each squad has 2 three-man teams. Total: 24 three-man teams.	1 platoon deploys with each division brigade and provides DS during the assault phase. Then platoons revert to GS.	During the assault phase, 6 mounted, mobile or fixed-position teams deploy to provide DS for each brigade and 6 teams deploy to provide GS near the division main CP OR after the assault phase, 1 division EPW collection point (6 teams) and 18 mounted, mobile patrols or fixed-position teams provide GS OR 24 mounted, mobile or fixed-position teams are used in any combination.	The division band to provide close-in security for the division CP or to guard EPWs AND the corps CS MP to conduct combat operations within the division rear and provide sustained MMS, area security, special operations support, and escort for evacuating EPWs/CIs between division collection points and to the corps holding area AND the corps L&O MP and CID to conduct sustained L&O operations and criminal investigations AND division/corps transportation assets for evacuation of EPWs **NOTE: Both heavy division companies require corps CS MP augmentation for each forward support platoon (1 corps MP squad per forward support platoon).**
MP Company (Light Infantry Division)	Perform the 5 MP functions.	A PM cell and 3 platoons. Each platoon has 3 squads and each squad has 2 three-man teams. Total: 18 three-man teams	Provide GS for all units in the AO.	Provide screening security for the division main CP (6 teams), the division EPW collection point (6 teams), and 6 mounted, mobile or fixed-position teams.	
MP Company (Heavy Division)* *19333F000	Perform the 5 MP functions.	A PM cell and 2 division support platoons. Each platoon has 3 squads and each squad has a squad leader and 3 three-man teams AND 3 forward support platoons. Each platoon has 2 squads and each squad has a squad leader and 3 three-man teams. Total: 36 three-man teams	2 platoons provide GS and 3 platoons provide DS.	Provide screening security for the division main CP (6 teams), the division EPW collection point (6 teams), and 6 mounted, mobile or fixed-position teams AND EITHER 3 DS platoons providing their brigade with either 1 forward EPW collection point (3 teams) and 3 mounted, mobile or fixed-position teams or 6 teams in any combination OR 18 mounted, mobile or fixed-position teams are used in any combination.	
MP Company (Heavy Division)* *19333L000	Perform the 5 MP functions.	A PM cell and 5 platoons. Each platoon has 2 squads and each squad has 3 three-man teams. Total: 36 three-man teams	3 platoons provide GS and 3 platoons provide DS.	Provide screening security for the division main CP (6 teams), the division EPW collection point (6 teams), and 6 mounted, mobile or fixed-position teams AND EITHER 3 DS platoons providing their brigade with either 1 forward EPW collection point (3 teams) and 3 mounted, mobile or fixed-position teams or 6 teams in any combination OR 18 mounted, mobile or fixed-position teams are used in any combination.	

Table 1-1. Unit Descriptions (continued)

MP Unit	Mission	Action Elements	Support Role	Capabilities (Include But Are Not Limited to [See TOEs for Details])	Augmentation Needed for Mission Accomplishment
MP Company (Air Assault Division)	Perform the 5 MP functions.	A PM cell and 4 platoons. Each platoon has 3 squads and each squad has 2 three-man teams. Total: 24 three-man teams	Provide GS for all units in the AO.	24 mounted, mobile or fixed-position teams are used in any combination.	(See augmentation needs on previous page.)
MP COMBAT SUPPORT UNITS					
HHC MP Brigade	Provide C2 and coordinate the combat, CS, and CSS of all MP elements assigned or attached.	Command and staff elements and long-range planning section	Provide command, control, planning, and supervision for up to 6 MP battalions and other assigned or attached subordinate HQ.	Provide continuous command, control, coordination, planning, and supervision for subordinate units.	None
HHD MP Battalion (Combat Support)	Provide C2 and coordinate the combat, CS, and CSS of all MP elements assigned or attached.	Command and staff elements plus a support element	Provide command, control, planning, and supervision for up to 6 MP companies and other assigned or attached subordinate HQ.	Provide continuous command, control, coordination, planning, and supervision for subordinate units.	None
MP Company (Combat Support)	Perform the 5 MP functions.	4 platoons with 3 squads. Each squad has a squad leader and 3 three-man teams. Total: 36 three-man teams	Provide GS for all units in the AO.	Provide security for a unified or combined HQ or higher HQ or 1 main CP and 1 tactical CP OR operate 1 EPW holding area (9 teams) and 27 mounted, mobile or fixed-position teams for use in any combination.	Corps band assets to guard EPWs AND corps transportation for evacuation of EPWs AND external MP to escort EPWs
MP Company (Combat Support)	Perform the 5 MP functions.	4 platoons with 3 squads. Each squad has a squad leader and 3 three-man teams. Total: 36 three-man teams	Provide GS for all units in the AO.	Provide security for a unified or combined HQ or higher HQ or 1 main CP and 1 tactical CP OR operate 1 EPW holding area (9 teams) and 27 mounted, mobile or fixed-position teams for use in any combination.	Corps band assets to guard EPWs AND corps transportation for evacuation of EPWs AND external MP to escort EPWs
MP Company Arctic Support Augmentation Detachment	Augment MP units when operating in an arctic environment.	1 mechanic (63B10)	Provide mobility support to MP units.	Provide support as designated by the commander.	None
MP Detachment (C2)	Provide command and administrative personnel for MP law enforcement teams.	19503LA has a platoon HQS; 3 individuals. 19503LD has a commander and support personnel; 8 individuals.	Provide L&O. 19503LA provides platoon-level C2. 19503LD provides company-level command, control, and support.	Commands and controls law enforcement teams, investigations teams, MWD teams, and physical-security teams.	None

Table 1-1. Unit Descriptions (continued)

MP Unit	Mission	Action Elements	Support Role	Capabilities (Include But Are Not Limited to [See TOEs for Details])	Augmentation Needed for Mission Accomplishment
MP Detachment (Old Guard)	Perform the 5 MP functions.	4 squads. Each squad has 3 three-man teams. Total: 12 teams	Support the HHC old guard battalion.	Provide support as directed by the commander.	None
MP LAW AND ORDER (L&O) UNITS					
MP HQ Team	Provide C2, personnel administration, and logistical support to attached L&O augmentation teams.	Commander and support personnel	Provide L&O.	Provide C2, personnel administration, and logistical support to attached L&O augmentation teams.	None
MP Operations Team	Provide mission supervision, staff planning, and technical supervision required to support the L&O mission.	L&O operations officer and operations staff	Provide L&O.	Provide overall mission supervision, planning, employment, and coordination of support. Provide technical supervision and administrative support of MPI, TAI, and force protection MP. Provide an evidence custodian/police intelligence NCO to be responsible for confiscated property used in criminal offenses and to receive and collect the initial analysis of criminal/operational information/intelligence.	None
MP Desk or Desk and Record Team	Serve as the primary control point for L&O-related incidents and operations initiated by MP patrols.	2 three-man teams	Provide L&O.	Receive/record complaints, dispatch investigative persons, maintain control of offenders/detained persons, maintain the status of investigations, prepare reports, and provide criminal data to the police intelligence NCO for analysis.	None
MP Traffic Accident Investigations Team	Provide the technical capability to investigate traffic accidents.	1 two- or three-man team: 19517AD00 has 1 two-man team. 19523LE00 has 1 three-man team.	Provide L&O.	Augment MP capabilities to enforce MSR regulations; man checkpoints, roadblocks, and dismount points; patrol traffic areas; and perform traffic escort duties.	The 19517AD00 accident investigation team is dependent on the supported unit for transportation.
MP Investigation Team	Provide the technical capability to investigate criminal incidents and conduct surveillance operations.	One, two, or three-man teams: 19517AE has 1 two-man team. 19533LA has 1 one-man team. 19533LB has 1 one-man team. 19533LC has 1 three-man team.	Provide L&O.	Investigate criminal incidents, conduct surveillance, work with HN military and civilian police, and collect criminal and operational intelligence. 19533LA00 provides supervision for the teams. 19533LB00 provides investigative support for populations of not less than 700 troops. 19533LC00 provides investigative support for populations of not less than 2,100 troops.	Dependent on the supported unit for transportation

Table 1-1. Unit Descriptions (continued)

MP Unit	Mission	Action Elements	Support Role	Capabilities (Include But Are Not Limited to [See TOEs for Details])	Augmentation Needed for Mission Accomplishment
MP Force Protection and Physical-Security Teams	Provide the expertise and technical capability to assist units in safeguarding personnel, equipment, and facilities.	One, two, or three-man teams: 19517AF has 1 two-man team. 19593LA has 1 three-man team. 19593LB has 1 one-man team.	Provide L&O and area security.	Conduct vulnerability assessments and identify and prioritize critical facilities and key terrain within the AO. Identify mission-essential activities that are vulnerable to criminal acts or disruptive activities. Ensure that activities are inspected to determine if safeguards are adequate. Assist base/base cluster commanders with the development of internal defense plans.	The 19517AF is dependent on the 19517AB operations team for transportation.
MP Detachment (Patrol Supervision Team)	Provide supervision for 3 MP teams.	1 MP NCO	Provide L&O.	Plan, direct, and supervise the employment of assigned or attached MP teams.	Dependent on the supported unit for transportation
MP Detachment (Motor and Dismounted Patrol Team)	Perform L&O operations.	1 three-man team	Provide L&O.	Protect designated personnel or facilities. Perform route and area reconnaissance, enforce MSR regulations, provide refugee and straggler control, and disseminate information. Perform peacekeeping operations to maintain surveillance over an area, observe activities, and report findings. Preserve or establish L&O. Apprehend absentees or deserters (US military personnel) in conjunction with civil law enforcement agencies.	The dismounted patrol team is dependent on the supported unit for transportation, when required.
MP Detachment (Registration Team)	Perform vehicle and firearm registration.	1 three-man team	Provide L&O.	Service up to 10,000 personnel for registration of individuals, vehicles, and firearms on a closed post or area. Issue credentials prescribing limits of circulation and privileges. Process up to 50 personnel daily for fingerprints and photography services.	Dependent on the supported unit for transportation
Senior Military Customs Inspector Supervisor	Perform technical supervision, staff planning, and coordination for customs inspection teams.	1 MP NCO	Provide L&O.	Supervise two senior military customs teams to ensure that personnel, equipment, and material meet customs, immigration, Department of Agriculture, and other federal agency requirements for units and personnel redeploying to the US.	None

Table 1-1. Unit Descriptions (continued)

MP Unit	Mission	Action Elements	Support Role	Capabilities (Include But Are Not Limited to [See TOEs for Details])	Augmentation Needed for Mission Accomplishment
Senior Military Customs Inspection Team	Conduct customs inspections and train and supervise redeploying unit personnel to augment US customs inspectors.	1 five-man team	Provide L&O.	Conduct inspections and advise redeploying units. Train selected individuals from the redeploying unit to augment US customs efforts. NOTE: One team supports one port of embarkation.	The redeploying unit undergoing customs inspection, to augment US Customs efforts. The number of personnel is proportional to the size of the unit.
MILITARY WORKING DOG (MWD) UNITS					
MWD Kennel Master Team	Supervise MWD teams.	One- or two-man team. 19537AA has 1 two-man team. 19583LG has 1 one-man team.	Provide L&O, area security, and I/R operations support.	Supervise, plan, and coordinate MWD operations and support requirements. Provide MWD handler/canine proficiency certification.	None
MWD Explosive, Narcotics, and Patrol Team	Detect explosives and controlled substances and search for, detect, and control personnel in support of crime scene searches. Provide personal protection, MOUT, health and welfare inspections, and I/R and customs operations.	One-man/one-dog team. The 19537A000-series TOEs have 3 teams each. The 19583L000-series TOEs have 1 team each.	Provide L&O, area security, and I/R support.	The 19537A-series dog teams are capable of providing a 24-hour explosive, narcotics, and tracking handler/dog requirement or three concurrent short-duration missions each requiring 1 handler/dog team. The duration of the missions will vary based on the climate, the environment, and the individual dog's ability. The 19583L-series dog teams provide short- duration missions based on the dog's ability. Explosive and narcotics teams provide MWD patrol support when not employed with explosive and narcotics detection.	None
MILITARY POLICE INTERNMENT/RESETTLEMENT (I/R) UNITS					
MP Command (I/R)	Provide command, control, staff planning, and supervision of I/R operations performed by all assigned or attached elements.	Command and staff elements	Provide I/R operations support.	Provide command, control, and staff planning for 2 or more MP brigades.	MP IRIC AND MP EPW evacuation detachment

Table 1-1. Unit Descriptions (continued)

MP Unit	Mission	Action Elements	Support Role	Capabilities (Include But Are Not Limited to [See TOEs for Details])	Augmentation Needed for Mission Accomplishment
MP Brigade (I/R)	Provide command, control, and staff planning for I/R operations performed by all assigned or attached MP elements.	Command and staff elements	Provide I/R operations support.	Provide command, control, and staff planning for 2 to 7 MP I/R battalions or up to 21 MP I/R battalions when augmented by 7 brigade liaison detachments (BLDs). Plan and provide staff supervision of I/R collection and evacuation operations. Coordinate with HN military territorial organizations and civilian police authorities, nongovernment organizations, private volunteer organizations, and US federal agencies on I/R matters. Provide coordination and support for out-of-theater evacuation of EPWs, if required. Support posthostilities operations.	MP IRIC AND MP EPW evacuation detachment AND MP I/R BLD (TOE 19453LD00) **NOTE: The BLD expands the staff planning and coordination capabilities on a ratio of 1 BLD to 3 MP I/R battalions.**
MP Battalion (I/R)	Provide command, staff planning, administration, and logistical support for the operation of an I/R facility.	Command and staff elements	Provide I/R operations support.	Operate an internment facility for either EPWs/CIs or DCs or a confinement facility for US prisoners. (Never more that one category at the same time.)	MP detachment (I/R) (EPW/CI) for EPW/CI/DC internment missions AND MP detachment (I/R) (confinement) for US prisoner confinement missions AND MP company (EG) for security of the evacuation and/or movement of US prisoners or EPWs/CIs/DCs AND MP company (guard) for prisoner guard services AND Corps/EAC transportation assets for movement of EPWs/CIs/DCs

Table 1-1. Unit Descriptions (continued)

MP Unit	Mission	Action Elements	Support Role	Capabilities (Include But Are Not Limited to [See TOEs for Details])	Augmentation Needed for Mission Accomplishment
MP Detachment (I/R) (EPW/CI)	Augment the MP battalion (I/R) to provide supervision, administration, combat health support, and logistical support for operating an internment facility interning EPWs/CIs or housing DCs.	2 compound control sections (2 three-man teams), a work project section (1 two-man team), a personnel section, and a supply section.	Provide I/R operations support.	Provide command, control, and supervision of up to 1,000 EPWs/CIs or 2,000 DCs. Provide supply and subsistence support (to include supervision of food preparation) for up to 1,000 EPWs/CIs or 2,000 DCs. Direct the activities related to assigning and supervising work projects for EPWs/CIs. Augment the MP I/R battalion in processing and maintaining records for up to 1,000 EPWs/CIs or 2,000 DCs.	MP company (guard) for prisoner guard services
MP Company (Escort Guard)	Provide supervisory and security personnel for evacuating and/ or moving EPWs/CIs.	4 platoons with 3 squads each. Each squad has a squad leader and 3 three-man teams. Total: 36 teams	Provide I/R operations support.	Provide security for the movement of the following numbers of POWs or CIs by the methods indicated: Marching - 1,000 to 1,500. Vehicle - 1,500 to 2,000. Rail - 2,000 to 3,000. Air - 2,125 to 2,406.	Corps/EAC transportation assets for movement of EPWs/CIs
MP Company (Guard)	Provide guards for EPWs/CIs or US prisoners, installations, and facilities.	3 platoons with 3 squads each. Each squad has a squad leader and 2 five-man teams Total: 18 teams.	Provide I/R operations support.	Provide security for a confinement facility containing up to 500 US military prisoners. Provide guards for securing 1 EPW/CI compound containing up to 2,000 EPWs/CIs or 4,000 DCs. Provide security guards for 3 railway terminals, each having up to 8 tracks. (Includes EPW/CI rail movement operations and protection of sensitive material within the terminal.) Provide guards for 1 military installation or facility up to 240,000 square yards in size, containing sensitive material. (This unit can provide guards for material transit.)	None
MP I/R Processing Squad	Process EPWs/ CIs.	1 squad leader and 1 eight-man processing squad	Provide I/R operations support. Augment the I/R battalion processing capability to meet operational requirements.	Provide processing capabilities of about 8 EPWs/CIs per hour.	None

Table 1-1. Unit Descriptions (continued)

MP Unit	Mission	Action Elements	Support Role	Capabilities (Include But Are Not Limited to [See TOEs for Details])	Augmentation Needed for Mission Accomplishment
I/R Brigade Liaison	Expand the MP brigade (I/R) or ASCC C2 capabilities.	Liaison officer and support staff	Provide I/R operations support.	When assigned to the MP I/R brigade, provide staff augmentation that expands the brigade's staff planning, coordination, and C2 capabilities for 3 MP I/R battalions. When assigned to the ASCC, provide I/R staff augmentation and a liaison link with allied/HN forces to ensure that the care and handling of US captured EPWs/CIs is in compliance with the Geneva Conventions.	None
I/R Camp Liaison Team	Provide continuous accountability of EPWs/CIs captured by US forces and transferred to an allied/HN for internment.	Liaison officer and support staff	Provide I/R operations support.	Provide advice, as requested, to commanders and staffs of allied/HN-operated internment facilities. Verify arrival, forward records, and provide continuous accountability for US captured EPWs/CIs interned in allied/HN facilities. Monitor prisoner treatment to ensure compliance with the Geneva Conventions. Receive/certify allied/HN requests for reimbursement of expenses associated with interning EPWs/CIs captured by US forces.	None
I/R Processing Liaison Team	Process and verify the transfer of EPWs/CIs captured by US forces to a allied/HN facility.	Liaison officer and support staff	Provide I/R operations support.	Provide processing and transfer capability for US-captured prisoners to an allied/HN EPW/CI facility.	None
MP Detachment (EPW Evacuation)	Provide evacuation support of EPWs.	EPW coordination officer and staff plus 10 one-man POD/POE teams and 3 two-man POE teams	Provide I/R operations support.	Coordinate administrative, logistical, and transportation support for up to 7 PODs and 6 POEs used for out-of-theater EPW evacuation and the escort guard personnel that support the movements. Coordinate for theater security at PODs/POEs and intermediate refuel points.	Responsible POD/POE security forces

Table 1-1. Unit Descriptions (continued)

MP Unit	Mission	Action Elements	Support Role	Capabilities (Include But Are Not Limited to [See TOEs for Details])	Augmentation Needed for Mission Accomplishment
MP Detachment (Internment Resettlement Information Cener [IRIC])	Provide a US central agency in each theater of war to receive, process, maintain, disseminate, and transmit the required information and data relating to EPWs, American POWs, CIs, and DCs within the theater.	Command and staff personnel	Provide I/R operations support. Serve as the single source for collection and storage of EPW/CI information in theater. Forward information to the National Prisoner of War Information Center at DA.	Collect, process, and disseminate (to authorized agencies) information regarding interned personnel detained in theater and those released to the custody of allied/HN authorities within theater. Receive, document, and disseminate to the theater commander information received regarding American POWs, CIs, and foreign nationals who are captured, missing, or otherwise detained. Receive, store, and dispose of personal property belonging to interned personnel who have died, escaped, or been repatriated and any property belonging to enemy soldiers killed in action that is not disposed of through grave registration channels.	None
MP Detachment (I/R) (Confinement)	Augment an MP battalion (I/R) in operating a confinement facility. Provide supervision, administration, combat health support, and logistical support of US military prisoners.	Corrections officer and confinement facility staff	Provide I/R operations support.	Provides a control team for the command, control, and supervision of US military prisoners and a disciplinary guard team to supervise custodial personnel for a maximum custody of 500 prisoners. Provide administration, health service, supply, and food service personnel to augment the battalion in providing these services for internees. (The food service personnel supervise and train US military prisoners working in the internee dining facility.)	MP battalion (I/R) for command, staff, planning, and operational support AND MP company (guard) for prisoner guard services

Table 1-1. Unit Descriptions (continued)

MP Unit	Mission	Action Elements	Support Role	Capabilities (Include But Are Not Limited to [See TOEs for Details])	Augmentation Needed for Mission Accomplishment
MP Confinement Facility Detachment	Provide C2, staff planning, administration, and logistical support for a confinement facility for US military prisoners.	Commander and facility staff	Provide I/R operations support.	Provide command, control, and administrative support for a confinement facility.	MP guard company to provide exterior perimeter security and required guard support to the confinement facility AND MP correctional team supervisor (TOE19553LE) to provide supervision of the guard force AND MP correctional team work supervisor (TOE 19553LF) to provide supervision of prisoners
MP Correctional Team Supervisor	Provide correctional supervision of prisoner work guard forces within a confinement facility.	2 two-man teams	Provide I/R operations support.	Provide correctional supervision to guard force personnel and US military prisoners within a modular-configured confinement facility.	None
MP Correctional Team Work Supervisor	Provide direct control and supervision of military prisoners within a confinement facility.	1 two-man team	Provide I/R operation support.	Provide correctional control, supervisory functions, and escort duties for military prisoners within a modular-configured confinement facility.	None
MILITARY POLICE CRIMINAL INVESTIGATION (CID) UNITS					
MP Group (CID)	Provide command, control, staff planning, and supervision for all CID elements within the theater.	Command and staff elements	Provide L&O.	Provide command, control, staff planning, and supervision over all CID elements within the theater. **NOTE: The Criminal Investigation Command is a stovepipe organization.**	None

Table 1-1. Unit Descriptions (continued)

MP Unit	Mission	Action Elements	Support Role	Capabilities (Include But Are Not Limited to [See TOEs for Details])	Augmentation Needed for Mission Accomplishment
MP Battalion (CID)	Provide command, control, staff planning, and supervision for all assigned or attached CID elements.	Command and staff elements	Provide L&O.	Provides command, control, staff planning, and supervision over all assigned and attached CID elements.	None
MP Detachment (CID) HQ Cell	Provide C2, evidence custody control, and investigative administration support. Coordinate for personnel administration and logistic support.	Commander and support staff	Provide L&O.	Provide C2 of assigned CID SA sections and/or teams performing all CID operations in their area of responsibility.	None
MP Detach-ment (CID)	Provide criminal investigative support to Army commanders at all echelons.	2-man teams. The number of teams varies by organization and is as follows: DSE heavy - 4 DSE light - 3 Section A - 4 Section B - 4 Supervisory team - 1 Senior team - 1 SA team - 1	Provide L&O.	Supervise and conduct criminal investigations.	When directed, MP CS, L&O, and MWD support

Chapter 2

Battle Command

Command is the authority a commander in military service lawfully exercises over subordinates by virtue of rank and assignment. Leaders possessing command authority strive to use it with firmness, care, and skill.

FM 101-5-1

Battle command is the exercise of command in an operation against a hostile, thinking opponent. Battle command includes visualizing the current state and the desired end state, then formulating concepts of operations to get from one state to the other at the least cost. In addition to visualizing and formulating concepts, battle command encompasses assigning missions; prioritizing and allocating resources; selecting the critical time and place to act; and knowing how and when to make adjustments in the fight. Battle command enables MP commanders to lead, prioritize, and allocate assets required in support of the Army commander. MP commanders must observe, orient, decide, and act on their decisions quickly. Information is the key element in the battle-command process; therefore, the commander must have accurate and timely information upon which to base his decisions.

OVERVIEW

2-1. The battle command of MP units is typically decentralized due to the nature of their CS functions, METT-TC, and the needs of the Army commander. This places the burden of sound, timely decision making to the lowest levels. MP leaders must develop a keen sense of situational awareness and visualization, and they must constantly track the actions of supported units.

BATTLEFIELD VISUALIZATION

2-2. The ability to visualize the battlefield is a critical element of battle command. Battlefield visualization is an essential leadership attribute and is critical to accomplishing the mission. It is learned and attained through training, practice, experience, technical and tactical knowledge, and available battle-command technologies. It results when the MP commander understands the higher commander's intent, his assigned mission, the enemy, and the friendly force's capabilities and limitations. See Appendix D for further information on command technologies.

2-3. Battlefield visualization includes the MP commander's view of what his forces will do and the resources needed to do the mission. He envisions a sequence of actions that will cause his MP forces to perform at the desired end state. Ultimately, the MP commander's battlefield vision evolves into his intent and helps him develop his concept of operations.

COMMANDER'S INTENT

2-4. The commander's intent is a key part of Army orders. It is a clear, concise statement of what the force must do to succeed with respect to the enemy, the terrain, and the desired end state. It provides the link between the mission and the concept of operations by stating key tasks. These tasks, along with the mission, are the basis for subordinates to exercise initiative when unanticipated opportunities arise or when the original concept of operations no longer applies. MP leaders at all echelons must ensure that the mission and the commander's intent are understood two echelons down (see FM 101-5).

CONCEPT OF OPERATIONS

2-5. The commander's intent does not include the method by which the MP units will accomplish the mission. This method is called the commander's concept of operations. It must—

- Convey the commander's vision of how to accomplish the mission in a manner that allows his subordinates maximum initiative.
- Build around intelligence gathering and the precise employment of MP resources.
- Provide the basis for task organization, scheme of maneuver, terrain organization, tasks to subordinates, and synchronization.

COMMAND AND CONTROL RELATIONSHIPS

2-6. MP units are assigned to, attached to, or placed under the operational control (OPCON) of MP or other units they support. OPCON is the authority to perform command functions over subordinate forces. This includes organizing and employing commands and forces, assigning tasks, designing objectives, and giving authoritative direction necessary to accomplish the mission. MP C^2 relationships may be changed briefly to provide better support for a specific operation or to meet the needs of the supported commander. MP units may be placed under the OPCON of another unit commander for short-term operations. The MP unit remains in this relationship only as long as it is needed for that operation.

> MP support to the Bosnian municipal elections consisted of one division and two corps MP companies. These MP assets, attached to Task Force (TF) Eagle, were task-organized from different sources. The division MP company and the PM cell were organic to TF Eagle's mechanized infantry division headquarters, but the two corps MP units were from US Army Forces Command (FORSCOM) MP battalions in CONUS.

SUPPORT RELATIONSHIPS

2-7. MP units on the battlefield provide two types of support—general support (GS) and direct support (DS). Corps and EAC MP units provide GS to their respective corps/EAC subordinate commands. Light, airborne, and air-assault MP companies provide GS to their respective divisions. Heavy-division MP companies provide GS to the division rear and DS to the division's subordinate brigades.

STAFF RELATIONSHIPS

2-8. The PM for each level of command is that command's advisor on MP combat, CS, and CSS operations. The PM—

- Advises the commander and staff about MP abilities/capabilities.
- Supervises the preparation of plans and dictates policies.
- Coordinates MP operations.
- Assists and supervises the interaction of supporting and supported units.
- Reviews current MP operations.
- Coordinates with allied forces and HN military and civil police.
- Ensures that MP plans and operations supporting the commander's tactical plan are carried out.
- Recommends when and where to concentrate the command's MP assets.
- Supervises or monitors MP support in the command's AO.

2-9. The PM works daily with the commander and staff officers who employ MP resources and whose AORs influence MP support. The PM works closely with the coordinating staff at the appropriate command level to coordinate MP support. He ensures that MP planning is practical and flexible, that plans are coordinated with staff sections and subordinate commands, and that plans reflect manpower and resources needed by MP. (This includes the need for C^2, fire support, equipment, and supplies. It also includes construction, communication, transportation, and aviation support.) As new information is received, the PM reviews, updates, and modifies the plans. He ensures that the echelon commander gets the necessary MP support.

2-10. In the absence of specific directions or orders, the PM plans the use of MP assets. He evaluates the current operations and projects the future courses of action (COAs). He bases his plans on assumptions consistent with the commander's intent and a thorough knowledge of the situation and mission. The PM considers—

- METT-TC.
- Current estimates developed by the intelligence preparation of the battlefield (IPB) and the police information assessment process (PIAP).
- The environment within the AO. This includes the climate, the terrain, and obstacles. It also includes the legal authority and status of the force; the width, depth, size, and location of built-up areas; and the attitudes and abilities of the local populace.

- The types of units operating in the area (to include joint, combined, multinational, and interagency units) and the missions and capabilities of these units. This knowledge is imperative to understand their capability to counter threats in their area.
- The specific missions of MP units in the area and the impact that rear-area security operations will have on the ability of these units to perform other functions.
- Personnel, vehicles, and equipment in the MP units.

2-11. Coordination and communication between the PM and Army commanders is essential. Such actions ensure timely and efficient MP support to all levels of command during any operation. The informal, technical chain of coordination is an open line of communication between PMs at different echelons. The informal chain of coordination fosters cooperation and help among the MP elements at each echelon. For instance, when the division PM needs more assets to accomplish added missions, he initiates coordination with the corps PM. If the corps PM can provide support, the division PM formalizes his request for assistance through the division Assistant Chief of Staff, G3 (Operations and Plans) (G3).

Chapter 3

The Threat

In the 40-odd years of the Cold War, in many locations around the world, the Army performed a deterrent role as part of the containment strategy. In other places, at other times, the Army fulfilled the Nation's expectation in operations too small to be called "wars," although no less dangerous. To the soldier on the ground, Operations Urgent Fury in Grenada and Just Cause in Panama were indistinguishable from combat operations of their forefathers. Operations Provide Comfort in Iraq and Restore Hope in Somalia, although peace operations, also proved to be dangerous.

FM 100-1

The end of the Cold War has reduced, but not eliminated, the most immediate threat to the security of the US and other western nations. However, the absence of a dominant, identifiable threat has produced a far more complex and confusing strategic environment than the one that was present during the Cold War. Forward-deployed and CONUS-based ARFOR and civilians are and will continue to be engaged in a range of military actions. These actions stem from deterring conflicts to conducting peacetime engagement operations to providing support to civil agencies at home and abroad.

OVERVIEW

3-1. During the past decade, the US has deployed forces in multiple operations that have included crisis response in combat situations as well as participation in noncombat activities. The Army's presence in South Korea, Saudi Arabia, and Kuwait and its deployments to Somalia, Haiti, Bosnia, and Kosovo are clear indicators that the military must be prepared to face not only the traditional threat, but also a nontraditional, nonecheloned enemy. To support Army commanders successfully, MP leaders must understand the nature and complexity of these threats and how they can potentially affect the desired strategic, operational, and tactical end states.

REAR-AREA AND SUSTAINMENT OPERATIONS

3-2. The rear area for any particular command is the area extending forward from its rear boundary to the rear of the area assigned to the next lower level of command. This area is provided primarily for the performance of support functions. Operations in the rear area assure freedom of action and continuity of operations, sustainment, and C^2. Sustainment operations are those that enable shaping and decisive operations by assuring freedom of action and continuity of operations, CSS, and C^2 (see FM 3-0). Sustainment operations include the following elements:

- CSS.
- Rear-area and base security.
- Movement control.
- Terrain management.
- Infrastructure development.

3-3. During the Cold War, the danger to rear areas included forces that would be deployed in support of major soviet-style operations. The adversaries using the soviet model could be expected to engage in intense combat activity in their enemy's rear area. Their forces were prepared to penetrate into the enemy's rear and to attack and destroy its reserve forces and rear-area installations. To protect the rear areas, the MP were among the first mobile fighting forces available to the battlefield commander and thus, a source of combat power. Today, the Army commander uses the MP's flexibility and their modular-force training, adaptability, and mobility to serve as a combat multiplier throughout his entire AO. During sustainment operations, the MP perform all functions to ensure freedom of maneuver in support of the overall operational effort.

3-4. Failure to protect our forces during sustainment operations normally results in failure of the entire operation. Sustainment operations determine how fast ARFOR reconstitute and how far they can exploit success. The likelihood of MP units encountering the enemy and engaging in direct combat (not only in the rear area, but also during sustainment operations) cannot be underestimated.

3-5. Threats to rear-area and sustainment operations exist throughout the full spectrum of military operations. These threats may be related or independently engaged, but their effects are frequently cumulative. Threats to rear-area and sustainment operations are usually theater-dependent and are not limited to those outlined in this manual. Joint Publication (JP) 3-10 further discusses the threat in the rear area. Although JP 3-10 defines the threat in the context of a JRA, MP leaders can expect the same level of activity anywhere that US forces are deployed.

RECEPTION, STAGING, ONWARD MOVEMENT, AND INTEGRATION OPERATIONS

3-6. Reception, staging, onward movement, and integration (RSOI) operations consist of essential and interrelated processes in the AO that transform arriving personnel and materiel into forces capable of meeting operational requirements. During RSOI operations, the threat encountered will depend mostly on the type of entry, the nature of the operation, and the enemy. During major contingencies, forces deploy from power-projection platforms within the US or forward bases. The PM must plan MP support during the initial stages of the deployment to ensure the protection of follow-on forces and the detection of potential threats (see FM 100-17-3).

3-7. MP support to RSOI operations includes, but is not limited to—

- Conducting AS operations to counter or prevent enemy actions against marshalling and staging areas.
- Conducting convoy, airport, and rail security operations.
- Conducting populace- and resource-control operations.
- Conducting other physical-security and force-protection measures.
- Conducting other MP functions (as determined by the PM).

THREAT LEVELS

3-8. The threat is divided into three levels. These levels provide a general description and categorization of threat activities, identify the defense requirements to counter them, and establish a common reference for planning guidelines. MP leaders must understand that this does not imply that threat activities will occur in a specific sequence or that there is a necessary interrelationship between each level.

LEVEL I

3-9. Level I threats include the following types of individuals or activities:

- **Enemy-controlled agents.** Enemy-controlled agents are a potential threat throughout the rear area. Their primary missions include espionage, sabotage, subversion, and criminal activities. Their activities span the range of military operations and may increase during both war and military operations other than war (MOOTW). These activities may include assassinating or kidnapping key military or civilian personnel or guiding special-purpose individuals or teams to targets in the rear area.
- **Enemy sympathizers.** Civilians sympathetic to the enemy may become significant threats to US and multinational operations. They may be the most difficult to neutralize because they are normally not part of an established enemy-agent network, and their actions will be random and unpredictable. During war and MOOTW, indigenous groups sympathetic to the enemy or those simply opposed to the US can be expected to provide assistance, information, and shelter to guerrilla and enemy unconventional or special-purpose forces operating in the rear area.
- **Terrorism.** Terrorists are among the most difficult threats to neutralize and destroy. Their actions span the full spectrum of military operations.
- **Civil disturbances.** Civil disturbances, such as demonstrations and riots, may pose a direct or indirect threat to military operations. Although this threat may not be of great impact during war, it may significantly change and affect MOOTW.

LEVEL II

3-10. Level II threats include the following types of forces:

- **Guerilla forces.** Irregular and predominantly indigenous forces conducting guerrilla warfare can pose a serious threat to military forces and civilians. They can cause significant disruptions to the orderly conduct of the local government and services.

- **Unconventional forces.** Special-operations forces (SOF) are highly trained in unconventional-warfare techniques. They are normally inserted surreptitiously into the rear area before the onset of an armed conflict. They establish and activate espionage networks, collect intelligence, carry out specific sabotage missions, develop target lists, and conduct damage assessments of targets struck.
- **Small tactical units.** Specially organized reconnaissance elements are capable of conducting raids and ambushes in addition to their primary reconnaissance and intelligence-gathering missions. Small (size or capability), bypassed conventional units, as well as other potential threat forces, are also capable of conducting raids and ambushes to disrupt operations.

LEVEL III

3-11. Level III threats are made up of conventional forces. Potential threat forces are capable of projecting combat power rapidly by land, air, or sea deep into the rear area. Specific examples include airborne, heliborne, and amphibious operations; large, combined-arms, ground-forces operations; and bypassed units and infiltration operations involving large numbers of individuals or small groups infiltrated into the rear area, regrouped at predetermined times and locations, and committed against priority targets. Level III forces may use a combination of the following tactics as a precursor to a full-scale offensive operation:

- **Air or missile attack.** Threat forces may be capable of launching an air or missile attack throughout the rear area. It is often difficult to distinguish quickly between a limited or full-scale attack before impact; therefore, protective measures will normally be based on the maximum threat capability.
- **Nuclear, biological, and chemical (NBC) attack.** Commanders must be aware that NBC munitions may be used in conjunction with air, missile, or other conventional-force attacks. The NBC weapons could also be used at Level I or II by terrorists or unconventional forces in order to accomplish their political or military objectives.

THREAT-LEVEL MATRIX

3-12. Table 3-1 lists the threat levels and their likely appropriate responses. The threat levels listed are based on the type of threat. The table should not be construed as restricting the response options to any particular threat.

THREAT PRIORITIES

3-13. The threat will attempt to perform the following operations against targets in the rear area:

- Detect and identify targets.
- Destroy or neutralize operational weapons-system capabilities.
- Delay or disrupt the timely movement of forces and supplies.

Table 3-1. Threat Levels

Threat Level	Example	Response
I	Agents, saboteurs, sympathizers, and terrorists	Unit, base, and base-cluster self-defense measures
II	Small tactical units, unconventional-warfare forces, guerrillas, and bypassed enemy forces	Self-defense measures and response forces with supporting fires
III	Large tactical-force operations (including airborne, heliborne, amphibious, infiltration, and bypassed enemy forces)	Timely commitment of a TCF

- Weaken the friendly force's C^2 network.
- Disrupt support to combat forces.
- Set the stage for future enemy operations.
- Create panic and confusion throughout the rear area.

3-14. Typical examples of enemy priority targets include the following:

- NBC-weapons storage sites and delivery systems.
- Key command, control, communications, computers, and intelligence (C4I) facilities.
- Air-defense artillery (ADA) sites.
- Airfields and air bases.
- Port facilities.
- Main supply routes (MSRs) and MSR checkpoints.
- Key LOC.
- Reserve assembly areas (AAs).
- Troop barracks.
- Critical civilian and logistics facilities.

THREAT LOCATION

3-15. The fact that the Cold War has ended does not imply that our traditional threat has ended. North Korea and Iraq are constant reminders of this fact. For the near future, Army commanders will fight units with Cold-War-era equipment and tactics. The Army trains and is prepared to fight an enemy capable of interfering with our freedom of maneuver throughout the battlefield. On an extended battlefield with asymmetric threats, the danger to high-value assets (HVAs) (including CSS, C^2, communication nodes, and MSRs) only increases. The idea that the danger to the rear area decreases as you travel farther away from the forward edge of the battle area (FEBA) is not true. Threat intensity does not depend on geographical location; it depends on what operations the enemy believes must be initiated (and to what degree) to achieve its objective in the rear area. Military commanders depend on the MP to delay and defeat threats in their AO with a mobile reaction force.

COMMUNICATIONS ZONE

3-16. The nature of the COMMZ will encourage Level I and II threats to concentrate along the LOC and other areas of military significance. MP units will encounter an enemy that is capable of disrupting operations throughout the COMMZ while employing terrorist activities, enemy-controlled agent activities, enemy sympathizers, and saboteurs. If the enemy is Level III capable, MP leaders must expect infiltrations and air, missile, or NBC attacks as a precursor to a major Level III operation.

CORPS REAR AREA

3-17. The activities in Levels I and II will be similar in composition and density as in the COMMZ, but they will target key corps units, key facilities, and corps sustainment capabilities. The threat activities, especially at smaller unit levels, may even precede hostilities. MP leaders must be alert and prepared to encounter unconventional forces conducting diversionary or sabotage operations and small combat units conducting raids, ambushes, or reconnaissance operations or collecting special warfare intelligence. With the fast tempo of offensive operations, MP leaders must also be alert and prepared to encounter bypassed forces that can disrupt operations in the corps rear area.

DIVISION REAR AREA

3-18. The division rear area (DRA) contains many types of CS and CSS units and conducts many complex operations. As in the COMMZ and the corps rear area, the full spectrum of Level I, II, and III activities may occur in the DRA. The main target will be the division's HVA (including key C^2 facilities; airfields; artillery, aviation, and air-defense assets; LOC; and essential CSS units). The threat may conduct diversionary attacks, sabotages, raids, ambushes, and reconnaissance operations to affect the commander's freedom of maneuver and the continuity of operations. Unlike corps MP, the likelihood of division MP encountering bypassed enemy forces is expected. Failure to delay or defeat these forces will impact division operations.

OTHER TYPES OF THREATS

3-19. As US forces are deployed throughout the world, they will have to face nontraditional, asymmetric threats (other than those listed in Table 3-1, page 3-5) that may be geographically specific. As part of situational awareness, and in coordination with military intelligence (MI) and CID personnel, MP leaders must evaluate and assess the impact of these threats in their AO. A TO is vulnerable to any or a combination of the following threats:

- National or international organized crime.
- Narcotics traffickers.
- Narcotics terrorists.
- Extremist groups.
- Paramilitary groups.
- Ethnic or religious disputes.
- Trade in illegal weapons or strategic materials.

3-20. MP leaders must be aware that other threats exist and that they have the same potential as the Level I and II threats to disrupt operations in rear-area or sustainment operations. In some instances, the above threats' capabilities or the massing of personnel may have the same potential threat as a Level III threat.

COUNTERING THE THREAT

3-21. The disruption of rear-area and sustainment operations directly affects military efforts. Three types of forces may be used to counter the threat in these areas—a base/base-cluster self-defense force, a response force, or a tactical combat force (TCF).

BASE/BASE-CLUSTER SELF-DEFENSE FORCE

3-22. A base cluster is established when the appropriate echelon rear-operations cell or command post (CP) places geographically contiguous or noncontiguous bases under the control of a headquarters. The base cluster becomes the next higher tactical C^2 headquarters of those bases. The rear-operations cell or the rear CP may also establish a base cluster for a corps support group (CSG), an area support group (ASG), or other CSS units operating in the corps or division rear areas.

3-23. US ARFOR have the inherent responsibility to contribute as many forces as possible for base defense and local security for themselves and their facilities, installations, and activities. Each base and base-cluster commander must develop a defense plan to detect, defeat, and minimize the effects of Level I and limited Level II threat attacks on his base or base cluster (including NBC attacks). To maximize the unit's mission accomplishment, defense plans must be flexible and allow for differing degrees of security based on the probability of threat activity. Defense plans are given to MP units operating near the base or base cluster. The base commander most often employs a series of defense measures providing internal and perimeter security. His internal reaction forces use organic weapons to neutralize and defeat most low-level threat activity. Although not fully equipped to engage major conventional or unconventional enemy forces that may confront him, a base commander must deploy his personnel to defend themselves until MP, HN, local police, or combat forces (if available) can respond.

RESPONSE FORCE

3-24. A response force is summoned when the base or base cluster is faced with threat forces that are beyond their self-defense capability. If the MP are the designated response force, they must—

- Coordinate with the supported bases or base-cluster commanders to conduct a joint IPB.
- Review base and base-cluster self-defense plans.
- Exchange signal-operating-instructions (SOI) information.
- Identify MP contingency plans to counter likely enemy activities.
- Integrate ADA, engineer, chemical, field-artillery (FA), Army-aviation, and close-air-support (CAS) fire support into their plans (if available).

3-25. MP units help the base or base cluster return to its primary mission by defeating Level II threats. MP units closely watch likely avenues of enemy approach, possible landing zones (LZs), drop zones (DZs), C^2 facilities, and other key installations. They accomplish this through the MMS, AS, and PIO functions.

3-26. If MP units are not the designated response force, they may become the initial response force for units within their AO. When this occurs, they block, delay or, if possible, destroy enemy elements within their capability. If the attack is by a larger or more capable force, they will maintain contact and continue to develop the situation or delay until the appropriate response force appears or the battlefield commander commits the TCF.

TACTICAL COMBAT FORCE

3-27. When the MP response force encounters or engages threats beyond its ability to defeat, it immediately notifies the higher headquarters. The battlefield commander will then evaluate the situation and commit the TCF to defeat the Level III threat. The TCF is normally a combined-arms organization tailored by the corps or division G3, based on METT-TC. The TCF normally receives fire, aviation, or other support needed to fight and defeat the threat. Once the TCF is identified and before it is committed to battle, it will conduct direct coordination with the MP or other response forces regarding the exchange of reconnaissance information, battle-handoff procedures, and contingency plans for TCF operations. Once the TCF is committed, the MP unit performing as the response force becomes OPCON to the TCF commander.

MP AS THE TCF

3-28. The MP brigade or battalion is capable of conducting TCF missions only when properly augmented. MP augmentation may be in the form of fires, small combat units, aviation assets, or CAS. The specific type of augmentation is METT-TC dependent. Additionally, the MP C^2 headquarters must receive the respective liaison officers to ensure that augmentation forces are synchronized and employed according to their capabilities. The MP commander's situational awareness and battlefield visualization are key elements to TCF operations. Once designated as the TCF, the MP unit commander establishes liaison with the appropriate rear CP to obtain—

- The current rear-area IPB.
- The friendly unit disposition.
- Defense plans.
- Priorities for protection.
- The fire-support plan.

3-29. Based on the above information, the MP commander conducts his own IPB and develops a concept of operations. He then forwards it to the appropriate higher echelon for coordination and approval.

3-30. The MP's ability to employ organic MP assets as part of the TCF is limited by the following factors:

- **MP availability.** Normally, all MP assets available are committed at all times. The specific function and scope that MP units perform during the operation are determined by the Army commander's needs, the intensity of the conflict, and the availability of MP resources. The commander, with advice from the MP leader, must decide which MP operations must be scaled back, delayed, or shifted before the MP unit can be designated as part of the TCF.
- **MP dispersion.** MP units are normally displaced over a large geographical area. Technological capabilities and mobility allow them to operate over great distances. In today's battlefield, a typical MP company employment covers between 1,000 and 1,200 square kilometers and performs numerous missions in support of all five functions. The distance between elements, the reprioritization and movement of other MP units, the difficult terrain, poor roads, and bad weather may slow down the MP's commitment as a TCF.

Chapter 4

Military Police Functions

Military police support the Army commander's mission to win the battle. They help the commander shape the battlefield so that he can conduct decisive operations to destroy enemy forces, large or small, wherever and whenever the Army is sent to war.

MAJ(P) Anthony Cruz

The MP Corps supports shaping and sustainment operations while performing its five basic functions as a flexible, economy-of-force organization. Through these functions, MP units are able to provide the commander with an array of CS operations across the full spectrum of military operations (see Table 4-1).

Table 4-1. MP Functions

	Functions				
	MMS	**AS**	**PIO**	**L&O**	**I/R**
Subtasks	Support to river-crossing and breaching operations and passage of lines Straggler and dislocated-civilian control Route R&S MSR regulation enforcement	Recon operations ADC Base/air-base defense Response-force/ TCF operations Critical site, asset, and HRP security Force protection/ physical security Antiterrorism	IPB support PIAP Active and passive roles Information collection and dissemination Joint, interagency, and multinational coordination	Law enforcement Criminal investigations US Customs operations Related L&O training	EPW/CI handling US military prisoner handling Populace and resource control Dislocated civilians
NOTE: Subtasks not all-inclusive.					

OVERVIEW

4-1. MP assets are limited. Specific functions are performed at any given time and are determined by the supported commander's need, the intensity of the conflict, and the availability of MP resources. The supported commander, through the command's PM, sets the priorities for MP operations.

4-2. The PM (based on METT-TC and the available assets) continuously evaluates the trade-off between the MP support that the commander requires and the MP support that can be provided. To meet the priorities set by the commander's tactical plan, the PM recommends the allocation and employment of MP assets for MP combat, CS, and CSS operations.

MANEUVER AND MOBILITY SUPPORT

4-3. The MMS function involves numerous measures and actions necessary to support the commander's freedom of movement in his AOR. The MP expedite the forward and lateral movement of combat resources and ensure that commanders get forces, supplies, and equipment when and where they are needed. This is particularly important in the modern battlefield where there is a greater geographical dispersal of forces and lengthened LOC.

4-4. The MP maintain the security and viability of the strategic and tactical LOC to ensure that the commander can deploy and employ his forces. The MP support the commander and help expedite military traffic by operating traffic-control posts (TCPs), defilades, or mobile patrols; erecting route signs on MSRs or alternate supply routes (ASRs); or conducting a reconnaissance for bypassed or additional routes. The MP move all units quickly and smoothly with the least amount of interference possible.

4-5. As part of the MMS function, the MP support river-crossing operations, breaching operations, and a passage of lines. They also provide straggler control, dislocated-civilian control, route reconnaissance and surveillance (R&S), and MSR regulation enforcement.

RIVER-CROSSING OPERATIONS

4-6. US forces conduct river-crossing operations to move a large force across a river obstacle with a minimum loss of momentum. The MP play a vital role by assisting the force commander in crossing the river as quickly and efficiently as possible. The crossing is usually planned and conducted by the headquarters directing the crossing. As such, a division crossing operation is conducted by a corps. Whether a brigade or division is crossing, the division MP company may also cross to provide uninterrupted support to the division. In these instances, there is a total reliance on the corps MP to support the crossing. The same is true for breaching operations and a passage of lines.

4-7. MP support for river-crossing operations reduces the crossing time and promotes the efficient movement of vehicles. It reduces congestion, speeds the crossing, and enables the maneuver commander to continue his momentum toward his primary objective. The MP establishes staging and holding areas and TCPs to control movement to and from these areas (according to the traffic-control plan). The MP may be called on to provide security for crossing forces at the crossing sites. In most cases, the MP TCPs and engineer regulation points (ERPs) are located on both sides of the river to improve communications and coordination between the units.

4-8. MP employment for river crossing is influenced by METT-TC. The number and placement of MP assets supporting a river-crossing operation varies with the size of the crossing force, the direction of the crossing (forward or retrograde), and the degree of enemy resistance expected or encountered.

4-9. The MP operating inside the crossing areas are OPCON to the crossing-area commander for the duration of the operation. The MP operating outside of the crossing area are under the command of their appropriate echelon commander.

4-10. The main thrust of MP support to river-crossing operations is within the immediate river-crossing site. The MP direct units to their proper locations (such as holding areas and staging areas) and ensure that units move through the area within the time listed on the movement schedule. This is a highly critical aspect of river crossing because the number of crossing sites is limited. MP assets are placed where they can stress MMS operations on MSRs leading into the crossing area.

4-11. The MP also provide AS to allow crossing forces to cross the river without losing momentum or forces. On both near and far sides, the MP are used to recon the crossing unit's flanks and rear to enhance security (see FM 19-4).

BREACHING OPERATIONS

4-12. Breaching operations are conducted to allow forces to maneuver through obstacles. Obstacle breaching is the employment of a combination of tactics and techniques to advance an attacking force to the farside of an obstacle that may be covered by fire. It is perhaps the most difficult combat task a force can encounter. Breaching operations begin when friendly forces detect an obstacle, and they end when the battle handover has occurred between the follow-on forces and a unit conducting the breaching operation (see FM 90-13-1).

4-13. The MP support breaching operations in numerous ways. MP assets are employed based on METT-TC, the available resources, and the commander's priorities. As a minimum, MP support may include, but is not limited to—

- Establishing TCPs along routes leading to or departing from the breaching site.
- Establishing holding areas.
- Establishing TCPs at the breaching site.
- Assisting engineers with temporary route signs.
- Establishing straggler-control operations.
- Conducting AS operations.

4-14. The most critical MP support is provided at the breaching site. The MP provide the commander with a means to control traffic flow to appropriate lanes. When multiple lanes branch off of a single far-recognition marker, the MP assist in directing the formation through various lanes. They also assist in modifying the traffic flow when lanes have been closed for maintenance or expansion. The MP conduct close coordination with the crossing-force commander and the TF commander executing the breaching operation. The MP enable the commander to make last-minute changes in traffic flow, thereby giving him increased flexibility to react to the enemy situation.

PASSAGE OF LINES

4-15. A passage of lines is an operation in which a force moves forward or rearward through another force's combat positions with the intention of moving into or out of contact with the enemy. The passage of lines is a high-risk military operation that requires close coordination between the passing unit, the stationary unit, and the MP providing the support.

4-16. The MP help reduce confusion and congestion during a passage of lines. They provide security in areas surrounding passage points and passage lanes to ensure that the passing unit has priority for using routes to and through the areas. The headquarters directing the operations sets the route's priority. The MP can support a forward, rearward, or lateral passage of lines. Before the actual operation, the MP in the AO conduct an area or zone reconnaissance to become familiar with the routes to, through, and beyond the area of passage. This enables the MP to extend the commander's C^2 by providing directions at passage points and by guiding the units through the passing lanes. Maintaining unit integrity and reducing incidents of stragglers is vital to maintaining the passing unit's momentum in a forward passage of lines. The MP perform aggressive straggler- and dislocated-civilian-control operations to prevent possible infiltration of the enemy.

4-17. A passage of lines is usually planned and coordinated by the headquarters directing the passage. A division's passage of lines is planned and coordinated by the corps headquarters. The detailed plans are made and coordinated between the divisions involved. Close coordination between the division and corps PMs is essential. An MP unit may be the unit involved in passing through the lines of another unit. When conducting a delay of a Level II threat, the MP are likely to conduct a passage of lines with the TCF. To avoid fratricide, close coordination between the MP response-force commander and the TCF is imperative (see FM 19-4).

STRAGGLER CONTROL

4-18. Mobile patrols, TCPs, and checkpoint teams return stragglers to military control as part of their operations. Most stragglers are simply persons who become separated from their command by events in the CZ or while moving through the COMMZ. If a straggler is ill, wounded, or in shock, an MP must give him first aid and, if needed, call for medical evacuation (MEDEVAC). If a straggler is uninjured, an MP directs him to his parent unit or to a replacement unit (as command policies dictate). The MP ensure that stragglers attempting to avoid return to their units are escorted back to their command (as a minimum).

4-19. The MP set up special posts for straggler control following NBC attacks or major enemy breakthroughs that result in large numbers of lost, dazed, and confused military personnel. Mobile MP teams operate between posts, and they also direct or collect stragglers. Straggler collection points may be needed if many stragglers are present in a combat theater. If allied forces are present in the theater, each nation establishes a collection point for its own personnel. MP teams are aware of each allied location and are prepared to assist allied soldiers in returning to their respective command. The MP use available transportation assets to transfer stragglers from TCPs and checkpoints to a straggler collection point. At the collection point, they are screened and sorted for removal to a medical facility or returned to their units to reconstitute the tactical commander's combat force.

4-20. The MP report information about stragglers with whom they come in contact. This information is compiled by the MP headquarters and forwarded through appropriate channels to the higher command. Information given by stragglers that is of immediate tactical value is reported without delay.

DISLOCATED-CIVILIAN CONTROL

4-21. The MP expediting traffic on MSRs may encounter dislocated civilians that could hinder military traffic. The MP assist and divert dislocated civilians from MSRs and other areas to I/R facilities. They may also deny the movement of civilians whose location, direction of movement, or actions may be a threat to themselves or to the combat operation. The HN government is responsible for identifying routes for the safe movement of dislocated civilians out of an AO. If needed, the MP assist the civil-affairs unit and the HN in redirecting dislocated civilians to alternate routes established by the HN government.

4-22. The US forces do not assume control of dislocated civilians unless requested to do so by the HN or unless operating in an environment with a hostile government. When the senior US commander assumes responsibility, the MP coordinate with civil affairs to set up TCPs at critical points along the route to direct dislocated civilians to secondary roadways and areas not used by military forces. As required, MWD teams may be used as a show of force or as a deterrent to assist with uncooperative personnel.

ROUTE R&S

4-23. The MP conduct hasty and deliberate route reconnaissances to obtain information on a route and nearby terrain from which the enemy can influence troop movement. A route reconnaissance focuses on continually monitoring the condition of MSRs, ASRs, and other areas. MP patrols look for restricting terrain, effects of weather on the route, damage to the route, NBC contamination, and enemy presence or absence. When enemy activity is spotted, the MP report it, maintain surveillance, and develop the situation. To gather information for proposed traffic plans, they look at the type and number of available routes; and they check load classifications, route widths, obstructions, and restrictions.

MSR REGULATION ENFORCEMENT

4-24. The MP undertake MSR regulation enforcement to keep the routes free for DSS operations. MP units support the command's MSR regulation measures as stated in the traffic-regulation plan (TRP). The TRP contains specific measures to ensure the smooth and efficient use of the road network. It assigns military route numbers and names, the direction of travel, highway regulation points, and preplanned MP TCPs. Most importantly to the MP, it gives the route's control classification. The MP ensure that classified routes are used only by authorized traffic. Vehicles traveling on roads too narrow for their passage or on roads unable to support their weight can obstruct the route.

4-25. To expedite traffic on MSRs, the MP operate special circulation control measures such as—

- Temporary route signing.
- Static posts such as TCPs, roadblocks, checkpoints, holding areas, and defilades at critical points.
- Mobile teams patrolling between static posts and monitoring traffic and road conditions.

4-26. They also gather information on friendly and enemy activities and help stranded vehicles and crews. The MP also place temporary route signs to warn of hazards or to guide drivers unfamiliar with the route. Using these measures, the MP exercise jurisdiction over the road network in the AO and coordinate with the HN (whenever possible) to expedite movement on MSRs.

AREA SECURITY

4-27. The MP perform the AS function to protect the force and to enhance the freedom of units to conduct their assigned missions. The MP who provide AS play a key role in supporting forces in rear-area and sustainment operations. The MP are a response force that delays and defeats enemy attempts to disrupt or demoralize military operations in the AO. The MP's mobility makes it possible for them to detect the threat as they aggressively patrol the AO, MSRs, key terrain, and critical assets. The MP's organic communications enable them to advise the appropriate headquarters, bases, base clusters, and moving units of impending enemy activities. With organic firepower, the MP are capable of engaging in decisive operations against a Level II threat and delaying (shaping) a Level III threat until commitment of the TCF.

4-28. Throughout all aspects of the AS function, the MP perform counteractions to protect the force and to prevent or defeat a Level II threat operating within the MP's AO. MP countermeasures may include implementing vulnerability assessments, developing procedures to detect terrorist actions before they occur, hardening likely targets, and conducting offensive operations to destroy the enemy. The MP use checkpoints and roadblocks to control the movement of vehicles, personnel, and materiel and to prevent illegal actions that may aid the enemy. The use of these control measures serves as a deterrence to terrorist activities, saboteurs, and other threats. However, at the same time, checkpoints and roadblocks expose the MP to these potential threats. To counter this fact, the MP may upgrade or harden vehicles and defensive positions.

4-29. The MP provide combat power to protect the C^2 headquarters, equipment, and services essential for mission success. The MP provide the battlefield commander with a light, mobile fighting force that can move, shoot, and communicate against any threat. Major subtasks associated with the AS function include reconnaissance operations; area damage control (ADC); base/air-base defense; response-force operations; and critical site, asset, and high-risk personnel (HRP) security.

RECONNAISSANCE OPERATIONS

4-30. As part of their AS mission, the MP serve as the eyes and ears of the battlefield commander by seeking out the enemy and reporting information obtained by recon patrols. The MP conduct area and zone reconnaissances, screening, surveillance, and countersurveillance to gain information to help guard against unexpected enemy attacks in the AO. The MP monitor likely avenues of approach and potential LZs and DZs. They become familiar with towns and other populated areas, ridgelines, woods, and other terrain features from which the enemy can influence movements along road networks. The MP pay close attention to areas near facilities designated critical by the commander. These areas include key MSR bridges and tunnels, depots, terminals, logistics-support bases, ammunition supply points (ASPs), communications centers/nodes, and C^2 headquarters. The MWD teams provide explosive detection and personnel detection/tracking capabilities that enhance reconnaissance operations (especially in urban terrain).

AREA DAMAGE CONTROL

4-31. MP units take measures to support ADC before, during, and after hostile actions or natural and man-made disasters. The ADC actions integrate CS and CSS functions for many units. Engineers, medical personnel, and Army aviators work closely to ensure quick relief operations. The MP provide MSR regulation enforcement, refugee control, and some local security when required. As with reconnaissance operations, the MP may use MWD explosive-and personnel-detection capabilities to augment all MP missions in rear-area and sustainment operations.

BASE/AIR-BASE DEFENSE

4-32. The MP are the base and base-cluster commanders' links for detection, early warning, and employment against enemy attacks. The information gathered is dispersed throughout the rear area to help apprise the commander of enemy activities near bases. Base defense is the cornerstone of rear-area security. When the threat exceeds the base/base-cluster capability, the base/ base-cluster commander requests MP assistance through the appropriate C^2 element.

4-33. Air-base defense requires special MP coordination with the US Air Force (USAF). The MP treat air bases like any other base or base cluster. A USAF air base may house the base-cluster commander, or it may be a cluster by itself. The MP are responsible for the air base's external defense. Its internal defense is primarily the responsibility of the Air Force's security forces. The security force provides in-depth defense for weapons, weapons systems, command centers, personnel, and other priority resources established by the base commander.

4-34. The security force is trained and equipped to detect, delay, and deny Level I and II threats. If a Level III threat is present, the security force is tasked with delaying actions; however, the HN, a sister service, or other support must be employed to defeat this threat. If the security force requires assistance to defeat a Level II threat, it may rely on MP response forces or another response force to assist in the defense. If available, the MP response force will react to the air-base defense, just as it would for any other base or base cluster within the MP's AO. However, the key to successful MP employment depends on the critical exchange of information before and during the MP employment. Good communications, an understanding of the defense plan, and liaison operations are vital in preventing responding forces from entering a situation that could result in fratricide.

RESPONSE-FORCE OPERATIONS

4-35. The MP are the base and base-cluster commanders' response force against enemy attacks in rear-area or sustainment operations. The MP gather information about the enemy while performing missions throughout the AO. This information provides commanders with enemy activity near bases. When needed, the MP provide a mobile response force to respond to bases under attack and to destroy the enemy. A base commander's defense plan is the cornerstone for protecting rear-area and sustainment operations. The base commander is responsible for defeating all Level I threats. When this threat exceeds his capabilities, he requests MP support. The MP located near bases or patrolling or conducting AS operations consolidate their forces, respond as quickly as possible, and conduct combat operations to destroy the enemy. If needed, the MP conduct a battle handover to the TCF.

4-36. MP forces performing as a response force are capable of conducting the following offensive operations:

- A movement to contact.
- A hasty ambush.
- A hasty attack.
- A delay.
- A call for fire.
- A repel attack against critical sites.
- A defense of critical sites.

4-37. To conduct these missions, the MP consolidate into squads or platoons to delay, defeat, or defend against the threat. See FMs 71-3 and 71-100 for more information on battle-handover operations.

CRITICAL SITE, ASSET, AND HRP SECURITY

4-38. The MP perform their AS function across the entire designated AO. When the MP provide security around a critical site or asset, they usually provide a mobile security screen, taking advantage of its weapons and communications platforms. This standoff protection detects and defeats the threat before it can move within direct-fire range of the facilities. The MP may be tasked to provide detail security to key facilities, assets, and personnel.

4-39. The MP provide security to major CPs and other facilities within the AO. Their employment maximizes mobility, lethality, and communications capabilities as a security screen. They may be required to establish local AS measures (such as checkpoints and listening/observation posts) to further protect these facilities. The MP provide internal access-control points to critical facilities, and they act as a response force. When the critical CP relocates, the MP provide in-transit security. Other types of critical site security include ASPs; deep-water ports; petroleum, oil, and lubricants (POL) terminals and pipelines; trains and railways; and air bases.

4-40. The MP may provide convoy security for top-priority units transporting especially critical supplies to combat forces. MP assets should be employed primarily on aggressive patrolling, route, area, and zone reconnaissance measures that would create a safe and secure environment for all types of vehicular and unit movement.

4-41. The MP and the CID provide protective services to designated key personnel by providing access control to restricted areas within CPs, providing in-transit security, or providing static security measures around the clock. The MP coordinate with the CID when close-in protection of key personnel is needed. The MP and the CID also provide training for personal-protection countermeasures. The MWD teams may be employed to enhance MP and CID detection and protection capabilities.

INTERNMENT AND RESETTLEMENT

4-42. The Army is the Department of Defense's (DOD's) executive agent for all EPW/CI operations. Additionally, the Army is DOD's executive agent for long-term confinement of US military prisoners. Within the Army and through the combatant commander, the MP are tasked with coordinating shelter, protection, accountability, and sustainment for EPWs/CIs. The I/R function addresses MP roles when dealing with EPWs/CIs, dislocated civilians, and US military prisoners.

4-43. The I/R function is of humane as well as tactical importance. In any conflict involving US forces, safe and humane treatment of EPWs/CIs is required by international law. Military actions on the modern battlefield will result in many EPWs/CIs. Entire units of enemy forces, separated and disorganized by the shock of intensive combat, may be captured. This can place a tremendous challenge on tactical forces and can significantly reduce the capturing unit's combat effectiveness. The MP support the battlefield commander by relieving him of the problem of handling EPWs/CIs with combat forces. The MP perform their I/R function of collecting, evacuating, and securing EPWs throughout the AO. In this process, the MP coordinate with MI to collect information that may be used in current or future operations.

4-44. Although the CS MP unit initially handles EPWs/CIs, modular MP (I/R) battalions with assigned MP guard companies and supporting MWD teams are equipped and trained to handle this mission for the long term. A properly configured modular MP (I/R) battalion can support, safeguard, account for, guard, and provide humane treatment for up to 4,000 EPWs/CIs; 8,000 dislocated civilians; or 1,500 US military prisoners.

EPW/CI HANDLING

4-45. The MP are tasked with collecting EPWs/CIs from combat units as far forward as possible. The MP operate collection points and holding areas to temporarily secure EPWs/CIs until they can be evacuated to the next higher echelon's holding area. The MP escort-guard company assigned to the MP brigade (I/R) evacuate the EPWs/CIs from the corps's holding area to the COMMZ's internment facilities. The MP safeguard and maintain accountability, protect, and provide humane treatment for all personnel under their care.

4-46. In a mature TO, MP (I/R) units process EPWs/CIs collected by MP teams and other units in the CZ. MP guard companies assigned to the MP (I/R) units guard EPWs/CIs at designated camps (see FM 19-40).

POPULACE AND RESOURCE CONTROL

4-47. Populace and resource control (PRC) denies adversaries or insurgents access to the general population and resources. The MP supports civil-affairs personnel and the tactical commander in planning and conducting PRC programs employed during all military operations. These programs may consist of curfews, movement restrictions, resettling dislocated civilians, licensing, ration control, regulation enforcement, amnesty programs, inspecting facilities, and guarding humanitarian-assistance distributions. The MP also direct dislocated civilians to resettlement camps where they are cared for while NGOs work to coordinate their relocation.

4-48. The MP's security capability, acceptability, and interface with the populace make them suitable as the primary forces of choice in these operations. The MP I/R units are specifically designed to fill this need (see FM 41-10).

US MILITARY PRISONERS CONFINEMENT

4-49. The MP detain, sustain, protect, and evacuate US military prisoners. Whenever possible, soldiers awaiting trial remain in their units, unless reasonable grounds exist to believe that they will not appear at the trial, the pretrial hearing, or the investigation or that they will engage in serious criminal misconduct. Under either of these two pretrial confinement instances, the commander must also reasonably believe that a less severe form of restraint (such as conditions of liberty, restriction in lieu of apprehension, or apprehension) are inadequate. When these circumstances exist and other legal requirements are met, US military personnel may be placed in pretrial confinement under the MP's direct control. Convicted military prisoners are moved as soon as possible to confinement facilities outside of the AO.

4-50. MP confinement operations parallel (but are separate from) the MP's EPW/CI operations. No member of the US armed forces may be placed in confinement in immediate association with an EPW or other foreign nationals who are not members of the US armed forces. A confinement facility is maintained within the TO only if distance or the lack of transportation requires such a facility. When military prisoners are retained in theater, temporary field detention facilities may be established in the CZ and a field confinement facility in the COMMZ (see FM 19-40 and AR 190-47).

LAW AND ORDER

4-51. The L&O function consists of those measures necessary to enforce laws, directives, and punitive regulations. The MP's L&O function extends the battlefield commander's C^2. The MP, in close coordination with the CID, work to suppress the chance for criminal behavior throughout the AO. By coordinating and maintaining liaison with other DOD, HN, joint, and multinational agencies, the MP at all levels coordinate actions to remove conditions that may promote crime or that have the potential to affect the combat force. Crime-prevention measures and selective enforcement measures are also performed as part of other functions. For example, the MP investigate traffic accidents and regulate traffic as part of the MMS function. The L&O function includes major areas such as law enforcement, criminal investigations, and support to US Customs operations. The primary units conducting L&O are the L&O detachments, customs teams, and CID units. Both the MWD team and the MP company (CS) also support the L&O function.

LAW ENFORCEMENT

4-52. Law-enforcement operations assist the battlefield commander in preserving his force. The MP dedicate assets to conduct law-enforcement operations based on the commander's needs. Since MP L&O and CID assets may be limited during the initial stages of any operation, the PM depends on the supported commander's development of an effective crime-prevention program and uses established investigative tools (such as inquiries and AR 15-6 investigations) to enforce rules. This will allow the PM to employ limited MP assets to perform other functions. Law-enforcement operations include responding to civil disturbances, conducting raids, investigating traffic accidents, conducting vehicle searches, supporting the commander's force-protection program, and providing support to HN and civil-enforcement agencies. Law enforcement also includes employing special-reaction teams (SRTs), marksman/observer (M/O) teams, and MWD teams.

4-53. The Army conducts counterdrug-support operations that generally fall within several DOD counterdrug-mission categories. The MP support the Army's role rather than directly participating in civil law-enforcement activities (such as searches, seizures, and arrests). When tasked, the MP provide training to law-enforcement agencies in common soldier skills, physical security, and tactical planning and operations. US Code (USC) Title 18, Posse Comitatus Act, Section 1385 prohibits the use of DOD assets to enforce civilian law (federal and state) except in cases and under circumstances expressly authorized by the Constitution or by an act of Congress.

4-54. In multinational operations, the MP may assist with the creation of multinational police units. Circumstances that may support the establishment of these police forces include existing or negotiated terms of international agreements or security-assistance programs, a multinational operational agreement, or appropriate military directives. The MP provide the capability to train foreign MP and/or reconstitute indigenous constabulary forces as part of stability and support operations. The MP can provide the initial mentoring to these forces and provide temporary, emergency law-enforcement capabilities until the foreign military or civilian police forces are functional (see FM 100-8).

CRIMINAL INVESTIGATIONS

4-55. The MP investigate offenses against US forces or property committed by persons subject to military law. Investigations against minor crimes (such as low-value, personal-property thefts or simple assaults) are normally investigated by the MP's L&O detachment. Investigations against major incidents involving wrongful death, serious bodily injury, and war crimes are referred to special agents of the USACIDC. The USACIDC conducts death investigations in the absence of HN agreements or in conjunction with the HN. The USACIDC special agents are authorized to investigate any alleged criminal conduct in which there is an Army interest unless prohibited by law or higher authority.

4-56. The USACIDC's investigative authority and investigative responsibility outside of the US are determined by international treaty or agreement (including status of forces agreements [SOFAs]), the policies of the HN government, the US ambassador, and AR 195-2. In the absence of such provisions, the following guidelines apply:

- On Army-controlled installations, the USACIDC has the authority to investigate alleged crimes.
- Outside of an Army-controlled installation, the USACIDC may investigate after coordinating with HN authorities.

4-57. In all environments, the USACIDC has the responsibility to investigate all felony crimes involving Army personnel, DA civilians and agencies, and companies working for the Army. The USACIDC investigates war crimes and crimes involving personal and government property affecting the Army's mission (logistics security [LOGSEC]). Other investigations (such as those based on international treaties, SOFAs, and joint investigations with the HN) may be undertaken if requested by the supported commander in support of the overall Army mission. See Chapter 9 for a complete discussion of the USACIDC.

US CUSTOMS OPERATIONS

4-58. The MP support the US Customs Service (USCS), the US Department of Agriculture (USDA), other federal agencies, joint staffs, and commanders who enforce the laws and regulations of the US concerning customs, agriculture, and immigration border clearances. Support to the USCS also includes assistance to federal agencies to eliminate the illegal traffic of controlled substances and other contraband through Army channels. MP support to customs operations are normally performed by specially trained MP customs teams. Although other MP units are not trained in all facets of customs operations, they may assist MP customs teams, the USCS or the USDA, and other federal agencies in the enforcement of applicable laws and regulations. When tasked, the MP/CID supports the investigation of violations of US Codes, DOD or DA regulations, and applicable provisions of SOFAs.

4-59. The MP report violations of customs laws, regulations, inspections, and investigative results to the installation's PM, the supported commander, and affected units. During redeployment from outside the continental US (OCONUS) to CONUS installations, the MP support the USCS or USDA efforts to ensure that personnel, equipment, and materiel meet customs, immigration, and agriculture requirements as stated by all applicable laws and regulations. As with other functions, MWD teams may be employed in support of customs operations for the detection of explosives or narcotics.

> During operation Just Cause, an MP platoon temporarily assumed the customs mission at the main terminal of the Torrijos International Airport located just outside of Panama City. Their mission supported the air evacuation of foreign-national civilians and the redeployment of some of the initial-entry US forces.

POLICE INTELLIGENCE OPERATIONS

4-60. The PIO function supports, enhances, and contributes to the commander's protection program, situational awareness, and battlefield visualization by portraying relevant threat information that may affect his operational and tactical environments. This threat information—whether it is police, criminal, or combat information—is gathered while conducting MP functions. The PIO function—

- Demonstrates the MP's/CID's capability to collect relevant threat information actively or passively.
- Ensures that all information collected while conducting MMS, AS, I/R, and L&O functions continues to be reported through the proper channels so that it can be analyzed by the Intelligence Officer (US Army) (S2) or the Assistant Chief of Staff, G2 (Intelligence) (G2) with support from the appropriate MP echelon.
- Coordinates with USACIDC elements to employ data developed by the USACIDC's programs. These programs include—
  - The Combating Terrorism Program as outlined in AR 525-13 and CIDR 195-1.
  - The Criminal Intelligence Program (CIP).
  - Personal-security vulnerability assessments (PSVAs).
  - A crime threat analysis.
  - Logistics-security threat assessments (LSTAs).

NOTE: The MP/CID must ensure that criminal information is released according to existing controls and restraints.

- Maintains constant liaison and communication with the higher echelon S2/G2; psychological-operations (PSYOP) units; HN police and other law-enforcement agencies; joint, combined, interagency, and multinational forces; the staff judge advocate (SJA); the CMOC; civil-affairs teams; and the force-protection officer.

4-61. The MP brigade commander, the battalion commander, and the PM are responsible for the PIO function. As such, each one must determine the best way to employ the available staff resources to monitor the execution of the PIO function within his command.

IPB PROCESS AND PIO FUNCTION

4-62. The IPB is a systematic, continuous process for analyzing the threat and environment in a specific geographic area. It is designed to support staff estimates and military decision making. Applying the IPB process helps the commander selectively apply and maximize his combat power at critical points in time and space on the battlefield by—

- Determining the threat's likely COA.
- Describing the environment the unit is operating within and the environmental effects on the unit.

4-63. The IPB process consists of the following four steps:

- **Define the battlefield environment.** The S2/G2 identifies the battlefield characteristics that will influence friendly and threat operations, establishes the limits of the area of interest (AOI), and identifies gaps in current intelligence holdings.
- **Describe the battlefield effects.** The S2/G2 evaluates the environmental effects with which both sides must contend. The S2/G2 identifies the limitations and opportunities that the environment offers on the potential operations of friendly and threat forces. This evaluation process focuses on the general capabilities of each force until COAs are developed later in the IPB process. This environmental assessment always includes an examination of terrain and weather, but it may also include discussions of characteristics of geography and infrastructure and their effects on friendly and threat operations.
- **Evaluate the threat.** The S2/G2 and his staff analyze the command's intelligence holdings to determine how the threat normally organizes for combat and conducts operations under similar circumstances. When facing a well-known threat, the S2/G2 can rely on historical databases and threat models. When operating against a new or less known threat, he may need to develop his intelligence databases and threat models concurrently.
- **Determine the threat's COA.** Given what the threat normally prefers to do and the effects of the specific environment in which he is operating, his likely objectives and the COAs available to him are determined. The S2/G2 develops enemy COA models that depict the threat's available COAs. He also prepares event templates and matrices that focus intelligence and identify which COA the enemy will execute (see FM 34-130).

4-64. Although the S2/G2 has the staff responsibility for the command's IPB, he is not the only one who conducts or needs to understand and use the IPB. Every Army commander and staff member must understand and apply the IPB process during the staff planning process. The MP on the battlefield are no exception. The MP employ the IPB process as their first step in developing and implementing the PIO function within their commands.

4-65. During the IPB process, the S2/G2 uses all available databases, intelligence sources/products, and related MI disciplines to analyze the threat and the environment. The PIO function supports this process by providing the S2/G2 with collected police, criminal, and combat information that can directly and significantly contribute to the success of the MI effort. In addition to the combat information, the PIO function provides additional information on possible criminal threats and COAs that may support the S2's/G2's IPB process and that can be used by the commander to upgrade the force-protection posture.

POLICE INFORMATION ASSESSMENT PROCESS

4-66. Like the S2/G2 uses the IPB process to continuously analyze the threat and the environment in a specific geographical area, MP leaders use the PIAP as a tool to continuously collect, organize, interpret, and gain access to police/criminal information in support of the IPB process. Criminal trends may have an impact on the tactical scenario, and the PIAP is a method used to consider this threat and its impact on friendly forces. MP leaders cannot use the PIAP as a substitute for the IPB process—the PIAP complements the IPB (see Appendix E).

PASSIVE MODE PIO

4-67. Every MP conducts the PIO function in a passive mode during their normal day-to-day operations and across the full spectrum of military operations. In the passive mode, PIO are not a stand-alone function; as such, they cannot be separated from other MP functions.

4-68. During the performance of MMS, AS, I/R, and L&O functions, the MP develop and exchange information with other organizations in the AO. The MP obtain information through contact with civilians, NGOs, IHOs, local and HN police, multinational police, and other security forces. If the MP receive, observe, or encounter information (police, criminal, or combat) while performing these functions, they will immediately submit a report to relay information up the chain of command. This report may be in the form of a size, activity, location, unit, time, and equipment (SALUTE) report; a spot report (SPOTREP); or another appropriate report. When the higher echelon (brigade, battalion, or PM) receives this information, it is simultaneously integrated into the ongoing IPB/PIAP and forwarded to the higher echelon S2/G2 (see Figure 4-1, page 4-16).

4-69. If police/criminal information is obtained, the MP—

- Update the previous police/criminal estimates provided to the S2/G2.
- Identify new or potential criminal threats or trends in the AO.
- Consider recommending that the supported commander upgrade the force-protection level.
- Notify adjacent units of the potential criminal threat that may affect their forces.
- Consider reprioritizing MP support to the identified threat area.
- Share information with HN/local police and other agencies.

4-70. If combat information is obtained, the MP—

- Forward the information to the higher headquarters S2/G2.
- Forward the information to the MP chain of command, integrate it into the MP's IPB process and, if necessary, take appropriate action.
- Notify the adjacent unit of the potential threat that may affect their forces.

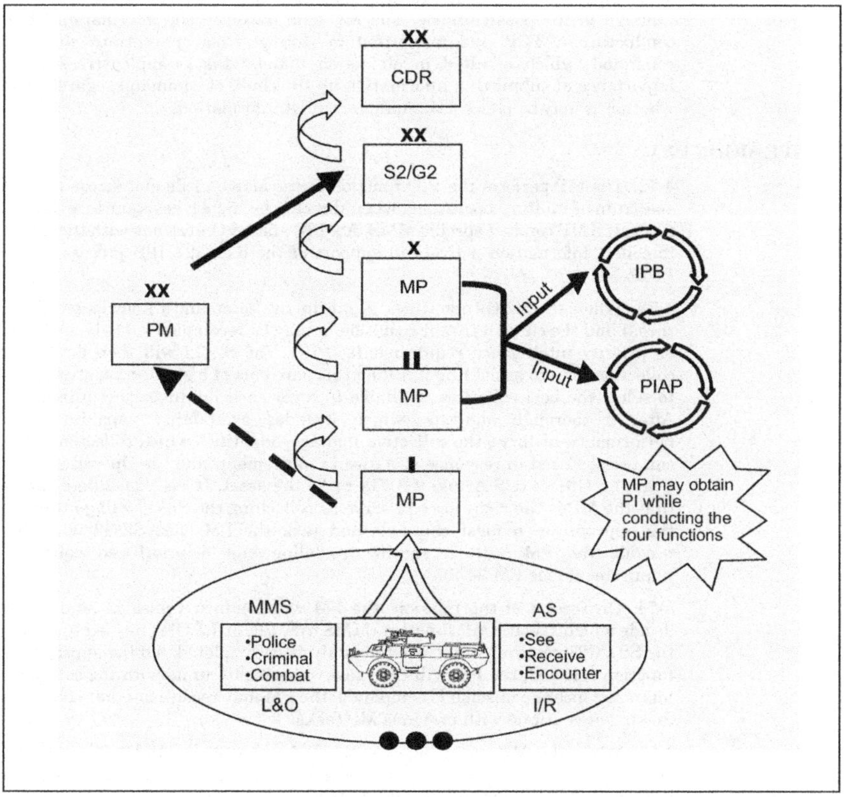

Figure 4-1. PIO—Passive Mode

> During Operation Uphold Democracy, an MP team was conducting a TCP as part of a cordon and search operation in Port-au-Prince, Haiti. While performing the task, two civilians approached the MP team informing them of criminal activity in the neighborhood. During the interview, the MP team prepared a sketch of a house and surrounding areas. The team also obtained information describing the criminals and their weapons. Recognizing that the criminal activity was in fact the actions of a political/mercenary group named FRAP, the MP team radioed the platoon leader and forwarded the field interview to higher headquarters. Two days later, a unit from the 10th Mountain Division raided the house, capturing weapons, ammunition, and equipment.

4-71. The preceding vignette demonstrates the MP performing the PIO function in the passive mode. The MP team received the information while conducting a TCP and submitted it through the appropriate chain of command, which resulted in an action taken. This example stresses the importance of submitting information up the chain of command regardless of whether it may be police, criminal, or combat information.

ACTIVE MODE PIO

4-72. The MP perform the PIO function in the active mode and across the full spectrum of military operations when directed by higher headquarters. In this mode, the MP conduct specific MMS, AS, I/R, and L&O missions with the intent to collect information actively in support of the S2's/G2's IPB process or the PIAP.

4-73. When the S2/G2 identifies a gap in the command's knowledge of the threat and the current threat situation, it may be recommended to be included as priority intelligence requirements (PIR). The S2/G2 will then develop a collection plan to assist him in filling this gap. Part of his collection strategy is to select the best collectors available to cover each intelligence requirement. After a thorough analysis (which includes availability, capability, and performance history), the collection manager identifies which collection asset can best be used in response to a given requirement, and the Operations and Training Officer (US Army) (S3)/G3 tasks the asset. If the S2/G2 determines that the MP is the right force to serve as collectors, the S3/G3 will go through the appropriate request channels and task the PM. The S2/G2 will then provide the PM with a specific guideline and a prioritized collection requirement (see FM 34-2).

4-74. On receipt of the mission, the PM will conduct a mission analysis to decide which specific MP function (MMS, AS, I/R, or L&O) is needed to satisfy the S2's/G2's requirement. Once the analysis is completed and the appropriate function selected, the PM will then task subordinate units with the collection mission. Once the mission is completed, the PM may receive another collection tasking or continue with previous MP tasks.

4-75. When the MP commander or the PM conducts the PIAP, he may also encounter a police/criminal information gap. This gap may become the MP commander's police/criminal information requirements (PCIR). If the gap cannot be filled with available data from the S2/G2, the CID, the HN, and other agencies, the MP commander/PM may task subordinate MP units or request CID assistance to support the collection effort. On receipt of the mission, the MP will then conduct a mission analysis and decide which specific MP function (MMS, AS, I/R, or L&O) is needed to satisfy the PM requirement. Once the mission is completed, the MP may receive another collection tasking or continue with previous MP tasks (see Figure 4-2, page 4-18).

4-76. Since any soldier on the battlefield can report police or criminal information, the MP commander or the PM must constantly coordinate and communicate with the S2/G2, PSYOP, and other agencies to obtain information that could be of MP/CID interest. This constant coordination is a key factor for ensuring that the MP/CID has visibility over the police/criminal information that is reported through non-MP channels.

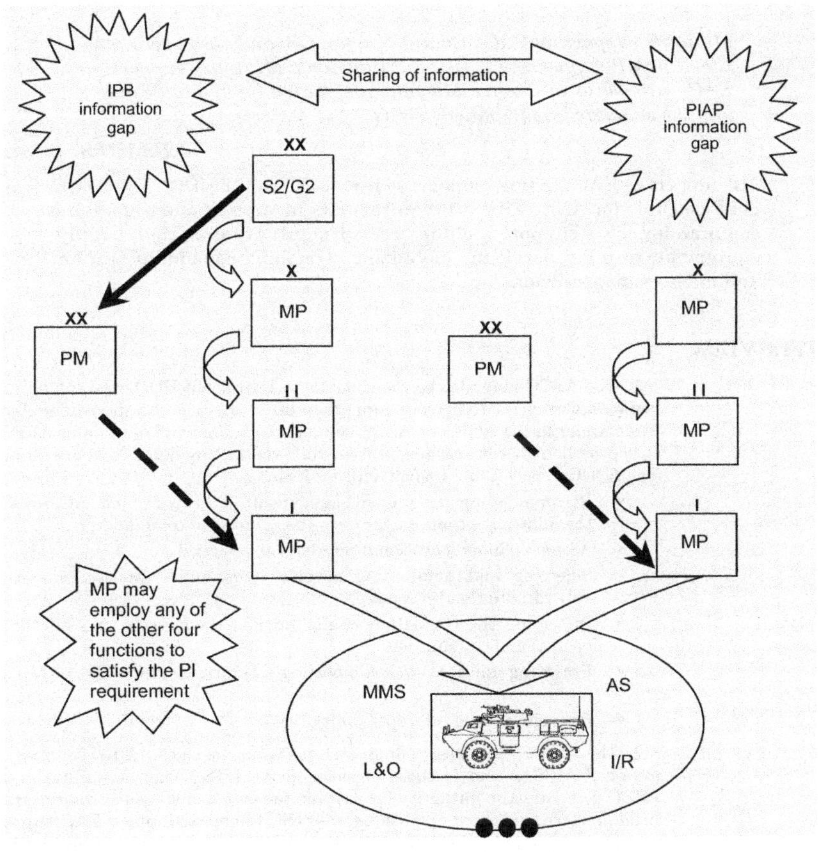

Figure 4-2. PIO—Active Mode

Chapter 5

Military Police Support to Echelons Above Corps

MP units supporting EAC perform combat, CS, and CSS operations. Like the MP supporting corps and divisions, MP units supporting EAC units perform the five MP functions based on available assets and the supported commander's needs.

USAMPS

MP support to EAC includes support to the ASCC and the TSC. The ASCC is responsible for Army Title 10 requirements in support of a combatant commander. This support includes recruitment, organization, supply, equipment, training, servicing, mobilizing, demobilizing, administration, and maintenance functions.

OVERVIEW

5-1. The ASCC may also be responsible for significant DOD- and combatant-commander-designated Army support to other services. As the senior Army commander in the AOR, the ASCC commander tailors and designates ARFOR to accomplish operational-level tasks while conducting major land operations. The ASCC's operational responsibilities include—

- Recommending the proper employment of Army-component forces to the joint-force commander or to the subunified commander.
- Accomplishing operational missions as assigned.
- Selecting and nominating specific Army units for assignment to subordinate theater forces.
- Informing the combatant commander of the Army's CSS effects on operational capabilities.
- Providing data to the supporting operations plans (OPLANs) as requested.
- Ensuring signal interoperability.

5-2. The ASCC provides administrative and logistics (A/L) services to assigned ARFOR and to those of subordinate JFCs. When appropriate, the ASCC delegates the authority for support tasks to a single subordinate Army headquarters. In major operations, the TSC (along with other EAC support commands) would be the ASCC's lead organization for planning, coordinating, executing, or providing required support functions (see FM 100-10).

5-3. The TSC is the senior Army support organization in a theater. Its commander reports to the ASCC or ARFOR commander. The TSC normally operates at the operational level of CSS with links to the strategic and tactical levels. Unity of command is the critical element that the TSC brings to the fight. The TSC is a multifunctional organization that centralizes the command, control, and supervision of support functions at EAC as directed by the ASCC or ARFOR commander. The TSC's mission is to maximize throughput and follow-on sustainment, including all CSS functions, of ARFOR and other designated supported elements. The TSC is capable of synchronizing logistics and other support operations for the ASCC. It provides area support to EAC units in the COMMZ and sustainment support to tactical forces. This support may include supply, procurement, property disposal, maintenance, transportation, field services, health services, civil-military affairs, MP support, engineer support, religious support, finance support, and personnel and administrative services.

5-4. Units and commands requiring support coordinate with the TSC support-operations staff to secure their initial support, to reestablish support, or to resolve support problems. In a fully developed theater, the TSC coordinates with a corps support command (COSCOM) for support of combat forces, although direct coordination with a division support command (DISCOM) is sometimes necessary. The TSC, augmented by a rear operations center, is also responsible for security operations as directed by the ASCC/ARFOR commander (see FM 63-4).

MILITARY POLICE SUPPORT

5-5. MP support to EAC units is provided through an array of multifunctional MP units. The nature of the operation, METT-TC, and the requirements of the supported commander will determine which type of MP unit is appropriate to augment, assign, attach, or place under OPCON to an EAC unit. The types of MP units that support EAC include CS, I/R, CID, and L&O teams (such as MWD or customs teams).

THE MP BRIGADE (CS)

5-6. The MP brigade (CS) provides MMS and AS to extended LOC within the COMMZ. These supply corridors include ports, inland waterways, railways, pipelines, airfields, MSRs, and MSR critical points. The MP support the users of the COMMZ's LOC by aggressively patrolling the area along the LOC. They play an important role in securing rear areas by performing combat operations against the threat. When properly augmented, the MP brigade headquarters may serve as the TSC's/ASCC's TCF headquarters. The MP provide MMS on the COMMZ MSRs leading into the corps's rear area. The MP implement the plans of HN and US staff elements to control the forward movement of combat resources along the LOC.

5-7. If resources are available, the MP brigade (CS) provides escorts to move US noncombatants (if present) from AA points to theater embarkation terminals. Until the MP brigade (I/R) arrives in theater, the CS MP units also perform EPW, confinement, and other operations normally performed by the MP I/R units.

5-8. The organization of an MP brigade (CS) supporting EAC includes the following:

- A brigade headquarters and headquarters company (HHC).
- Up to six MP battalions (each with up to six companies).

- Numerous L&O detachments and MWD teams.

5-9. Additionally, the ASCC's PM or commander may attach or direct OPCON of customs teams to the MP brigade (CS). Battalion and company organization in the MP brigade is the same as that in the corps MP brigade; however, METT-TC determines the number of battalions and companies. The MP brigade (CS) has additional MP companies to provide security for EAC-specific units/activities such as—

- A unified command.
- An ASCC and TSC headquarters.
- LOC seaports, airfields, and railways.
- EAC ammunition storage areas.
- EAC petroleum terminals and pipelines.

5-10. While the corps MP brigade (CS) MWD teams are employed in a GS role, the MP brigade (CS) MWD teams are employed to augment seaport security and to conduct inspections of postal items to detect explosive materiel and narcotics.

5-11. The C^2 within the MP brigade (CS) is consistent with that in any Army brigade. The MP brigade commander works directly for the EAC commander, the battalion commander works for the brigade commander, and the company commander works for the battalion commander. However, this usual C^2 relationship may be altered briefly (based on METT-TC) to enhance the overall EAC combat capability for responding to a Level II threat. For example, MP units operating within an ASG's AO may be under the OPCON of the ASG's rear-area operations center (RAOC), which directly tasks MP units responding to Level II threats. The same is true for placing MP units under the OPCON of the EAC's TCF headquarters for responding to Level III threats. Any conflict in mission priorities is resolved through MP C^2 channels.

5-12. The MP brigade commander is both the MP brigade commander and the EAC's PM. He employs his assets according to METT-TC and the commander's concept of operations. Factors affecting his employment of MP assets include the—

- Nature of the operation (joint, combined, or multinational).
- HN's ability to provide MP-related support (such as port security).
- Custody and location of EPWs/CIs during internment operations until I/R units arrive in theater.
- Number of kilometers of the MSR in relationship to movement-control requirements.
- Number and kinds of critical facilities.
- Number of HRP requiring close-in security.
- HN's ability to control the civilian populace, refugees, and dislocated civilians.
- Supply distribution strategy.
- Risk acceptance and threat in the AO.
- Communications requirements (such as using teams as relays).

5-13. Whenever possible, the MP brigade's AO coincides with the territorial responsibility of the supported command. The MP brigade commander assigns

5-17. In a mature theater in which large numbers of EPWs are captured, the EPWs' requirements may exceed the capacity of the MP brigade (I/R). In this instance, an I/R command is established. An MP command (I/R) has two or more MP brigades (I/R) and will normally be assigned to the ASCC. When the MP command (I/R) assumes OPCON of the MP brigades (I/R) from the lower echelon, it assumes that echelon's I/R mission.

5-18. If the US decides to transfer captured EPWs/CIs to the HN or to another nation, the US must ensure that the nation is a party to the Geneva convention and is willing and able to comply with the convention. In this case, the number and type of I/R MP units required for processing and retaining EPWs/CIs before the transfer is based on agreements and on METT-TC. Additionally, the MP brigade (I/R) is assigned I/R teams that are located at the processing and transfer points and at the HN or third-country EPW camps. The MP brigade (I/R) liaison team will supervise these dispersed teams to ensure that the HN or the third country provides adequate care and security of US-captured EPWs/CIs and that accountability is maintained according to the Geneva convention.

5-19. The MP brigade (I/R) HHC is the C^2 element for the brigade's assets. It consists of the following elements:

- A brigade command section.
- A company headquarters.
- An Adjutant (US Army) (S1).
- An S2.
- An S3.
- A Supply Officer (US Army) (S4).
- A Civil-Affairs Officer (US Army) (S5).
- A Communications Officer (US Army) (S6).
- Finance and accounting.
- Medical operations.
- Public affairs.
- A unit ministry team.
- An SJA.
- An inspector-general (IG) section.

5-20. Other brigade assets may include the following:

- An MP I/R BLD.
- An I/R information center.
- An MP escort-guard company.
- An MP I/R battalion headquarters.
- MP I/R (EPW/CI) detachments.
- MP I/R (confinement) detachments.
- MP guard companies.
- MWD teams.
- Processing squads, processing liaison teams, camp liaison teams, and evacuation teams (all as required to support EPW transfer or to conduct an out-of-theater evacuation).

the MP battalion's AO by the above factors as well as by METT-TC. For example, the AOR for an MP battalion may be a large population center of larger geographical areas in which CSS complexes and MSRs are located. But as employment factors and the commander's needs change, so will the MP's AOR. The MP brigade commander must move and tailor his forces to meet the current and projected mission requirements. Unlike many other EAC assets, MP units require 100 percent mobility to shift AOs frequently and rapidly. The following vignette depicts the required MP flexibility to support EAC operations:

> During Operation Desert Shield/Storm, MP units were supporting and moving units throughout their AOs at a such an extraordinary rate that many of them had to relocate their headquarters multiple times just to keep pace with their changing AO.

5-14. Most EAC MP assets are employed along LOC and around areas of high troop concentration. Few EAC MP are dedicated to support fixed commitments (such as ports, air bases, and headquarters [discussed in paragraph 5-8]). When supporting fixed commitments, MP units provide a mobile security screen, and they man static positions when directed or when based on METT-TC. The MP brigade (CS) may have to plan for and actually perform the evacuation and internment of EPWs/CIs and the confinement of US military prisoners until the MP brigade (I/R) arrives.

THE MP BRIGADE (I/R)

5-15. US policy requires that all persons held in Army custody be accorded humane care and treatment from the moment of custody to their final release. The policy applies to detained or interned civilians as well as to EPWs and confined US military personnel. This policy is equally binding on all US troops (see FM 19-40).

5-16. The ASCC or the TSC supports US laws, regulations, policies, and international agreements by providing personnel, administrative, morale, internment, resettlement, and confinement services for the TO. The TSC's MP brigade (I/R) in the TO provides this support. However, since most I/R units are in the reserve components, the initial I/R operations (as mentioned above) may have to be conducted by the MP brigade (CS). Once the I/R unit arrives in the AO, it is responsible for—

- Providing firm but humane control of EPWs/CIs and dislocated civilians.
- Coordinating with HN personnel, military territorial organizations, civilian police authorities, NGOs, private volunteer organizations, and US federal agencies on matters pertaining to I/R operations.
- Performing C^2 operations for all I/R units.
- Controlling, employing, and releasing EPWs/CIs as set forth by the Geneva convention and other international laws and by the UN and other governmental bodies.
- Handling US military prisoners.

5-21. The MP brigade (I/R) subordinate units are employed most often in the COMMZ near CSS facilities and are placed near sea, air, and rail terminals. They receive, process, and intern EPWs/CIs; confine US military prisoners; or assist in the resettlement of refugees or dislocated civilians.

5-22. The MP brigade (I/R) escort-guard company supports the evacuation of EPWs/CIs from the CZ. The company has a company headquarters and four platoons. The escort-guard company requires sufficient mobility to transport MP personnel to the CZ and to operate mobile teams while escorting the EPWs/CIs. The MP go forward to the corps's holding area to take custody of the EPWs/CIs. They may go forward to division collecting points, if distances and resources permit. Using any means of available transportation, the MP ensure that the EPWs/CIs are quickly evacuated to MP battalions (I/R) in the COMMZ. Close coordination with the EAC and corps movement-control centers and the corps MP brigade is required to ensure that transportation assets returning to the COMMZ are employed to evacuate EPWs/CIs from the corps's holding area. Walking wounded EPWs/CIs are evacuated by the same means as other EPWs/CIs, while litter patients are evacuated through medical channels. Guarding EPWs/CIs while in the MEDEVAC channels and during their hospitalization is not an MP mission; therefore, there is not an MP force structure to support this mission. In most instances, the impact of having the MP perform this mission causes trade-offs in missions for which they are responsible.

5-23. The theater MP brigade (I/R) and out-of-theater MP brigade (I/R) subordinate units will evacuate EPWs to internment sites within CONUS (if directed). The theater brigade structure is based on the projected capture rate over time and available out-of-theater transportation assets (frequency and capacity). The out-of-theater brigade structure is based on the total EPW/CI population, the number of internment sites, transportation nodes, and escort requirements. Theater escort-guard MP move the EPWs/CIs to the seaport and aerial port of embarkation (SPOE/APOE). The escort-guard MP assigned to the out-of-theater brigade escort the EPWs/CIs from the theater ports of embarkation (POEs) to the out-of-theater internment sites. The out-of-theater brigade is assigned an I/R evacuation detachment, which is employed at and coordinates the evacuation from the theater POEs, through the out-of-theater ports of debarkation (PODs), to the out-of-theater facilities.

THE MP BATTALION (I/R)

5-24. The MP battalion (I/R) is a modular organization and can be configured to operate internment facilities for EPWs/CIs, confine US military prisoners, or resettle dislocated civilians. When performing EPW/CI internment operations, the MP brigade (I/R) has up to 7 MP battalions (I/R); when augmented with the appropriate number of BLDs, it has up to 21 MP battalions (I/R). The ASCC, the TSC, and the MP brigade (I/R) must consider that the requirement to establish an MP battalion (I/R) internment facility is resource intensive. Therefore, MP I/R units, other supporting units, supplies, and equipment for the EPWs/CIs should arrive in theater ahead of the projected EPW/CI arrival at the internment facilities. Early arrival should be based on the time required to establish fully operational facilities (construct and man) and resupply operations before the EPWs/CIs arrive.

5-25. The MP battalion (I/R) has a command section, a company headquarters, and various staff sections. The staff sections provide the core battalion-level capabilities to conduct internment operations. The modulated design expands as the EPW/CI population increases. The battalion is assigned up to four detachments, two guard companies and, if needed, a processing squad. When fully operational, an MP battalion (I/R) operates an enclosure capable of interning 4,000 EPWs/CIs. The battalion mission centers on eight 500-man compounds. The battalion operates the compounds in close proximity to maximize its resources for the security and internment of the EPWs/CIs. This includes the resources needed to employ EPWs/CIs as a labor force according to the provisions of the Geneva convention.

5-26. Each MP detachment (I/R) (EPW/CI) operates two 500-man compounds and provides augmentation to the battalion staff sections to support 1,000 EPWs/CIs. Each guard company is capable of providing security for 2,000 EPWs/CIs. The guard company has a company headquarters and three platoons. The guard company requires sufficient mobility and communications to support routine battalion missions. While minimum mobility and communications is required to support EPW/CI internment operations, on-site guard personnel must often move considerable distances guarding labor groups performing work projects throughout the COMMZ.

5-27. The MP (EPW/CI) processing squad is capable of processing eight EPWs/CIs per hour and includes interpreters to support the processing. If processing squads are required to augment MP battalions (I/R), the operational requirements will be based on METT-TC.

MP-Battalion Resettlement Operations

5-28. The basic organization used for EPW/CI internment is used for resettlement operations. The primary mission-focus change is from guarding EPWs/CIs to protecting and controlling dislocated civilians. As such, an MP battalion (I/R) with four detachments and two guard companies is capable of supporting 8,000 dislocated civilians. However, the MP battalion (I/R) may require augmentation to conduct L&O operations associated with the resettlement. Augmentation may include the full scope of PM functions (operations, investigations, physical security, MP-station operations, and patrols) and civil affairs.

MP-Battalion Confinement Operations

5-29. When configured with the MP detachment (I/R) (confinement), the MP battalion (I/R) is capable of confining US military prisoners. The MP detachment (I/R) (confinement) provides trained corrections and support personnel required for confinement operations. As with the EPW/CI configuration, the modular confinement structure expands as the US prisoner population increases. With three confinement detachments and three guard companies, the MP battalion (I/R) has a maximum capacity of handling 1,500 US prisoners. Generally, only one MP brigade battalion (I/R) is configured for confinement, but the actual number will depend on the number of US prisoners requiring confinement within a theater. While theater policy for confinement operations remains with the ASCC commander, it is the MP brigade (I/R) that executes the mission.

5-30. When possible, soldiers awaiting trial remain in their units unless reasonable grounds exist to believe that they will not appear at the trial, the pretrial hearing, or the investigation or that they will engage in serious criminal misconduct. Under either of these two pretrial confinement instances, the commander must also reasonably believe that a less severe form of restraint (such as conditions of liberty, restriction in lieu of apprehension, or apprehension) is inadequate. When these circumstances exist and other legal requirements are met, US military personnel may be placed in pretrial confinement under the MP's direct control. Commanders may choose to establish field confinement facilities within their AO. However, corps and division MP companies have the expertise to operate only a field detention facility for a limited period of time. These units cannot operate a confinement facility and have neither the resources nor the capability to operate such a facility on an extended basis. Therefore, all confinement is consolidated in the COMMZ whenever possible.

5-31. All assets of the MP battalion (I/R) (confinement) are employed to detain, confine, sustain, and protect US prisoners. As with the battalions conducting EPW/CI and resettlement operations, the battalion conducting confinement operations is generally located in the rear of the COMMZ, near logistics and transportation support. This allows US prisoners to be moved as quickly as possible from the corps's area to the COMMZ's confinement facility. Movements of US prisoners from the COMMZ to CONUS will be according to DA policy.

Chapter 6

Military Police Support to Corps

Corps MP support their command by performing the MP functions critical to the success of their tactical commander's concept of operations.

USAMPS

Corps are the largest tactical units in the US Army. They are the instruments by which higher echelons conduct operations at the operational level. Higher headquarters tailor corps for the theater and the mission for which they are deployed. They contain organic combat, CS, and CSS capabilities to sustain operations for a considerable period of time. Corps are capable of operating in a joint and multinational environment, providing C^2 for up to five divisions and covering up to 35,000 square kilometers.

OVERVIEW

6-1. MP support to a corps is provided by an MP brigade (CS) assigned to each corps. The MP brigade provides combat, CS, and CSS throughout the corps's AO. However, subordinate MP units are not assigned to subordinate corps units. Instead, the MP brigade commander gives them an AO based on the corps commander's concept of operations. When possible, MP battalion AOs coincide with those of the CSG RAOCs.

6-2. The corps MP provide combat power within the command's rear area. They perform combat operations to counter Level II forces and to support the defeat of Level III forces. When properly augmented, the MP brigade may serve as the corps's TCF. The corps MP also provide a critical link between MP operations in the division and in the COMMZ. The corps MP support division commanders by helping the division MP conduct sustainment operations. The corps MP coordinate with the division MP for mutual support.

THE MILITARY POLICE BRIGADE (COMBAT SUPPORT)

6-3. The MP brigade (CS) supporting a corps contains a brigade headquarters, up to six MP battalions (CS), numerous L&O detachments, and MWD teams. Each MP battalion (CS) has up to six MP companies (CS). As with the EAC's MP brigade (CS), the number of battalions and companies is determined by METT-TC. The corps MP brigade (CS) has additional companies to support each division and to provide security for the corps and COSCOM headquarters and corps ammunition storage areas (see Figure 6-1, page 6-2).

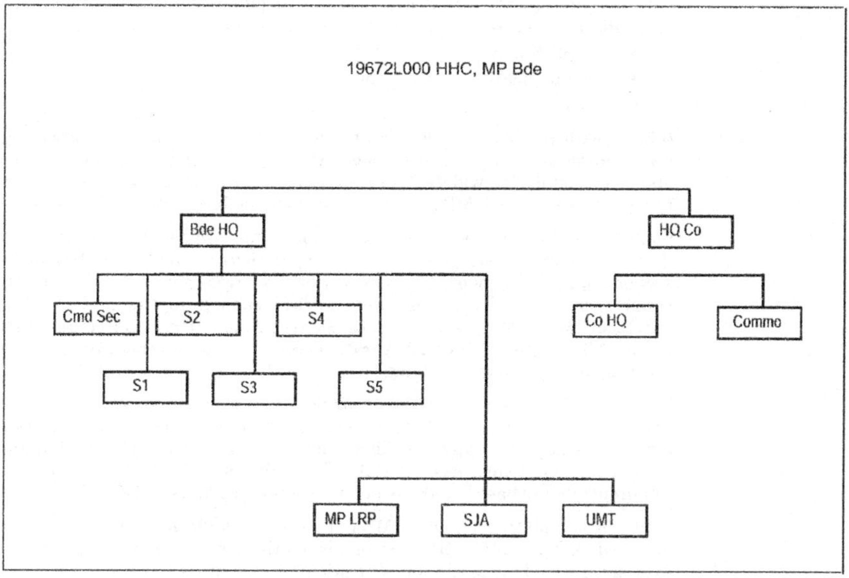

Figure 6-1. Military Police and USACIDC Organizations

6-4. The MP brigade HHC provides C^2 and A/L support to the brigade. The brigade HHC consists of a company headquarters and a brigade headquarters that contains the commander's immediate staff. The staff officers supervise the brigade's major organizational elements, including the—

- S1, S2, S3, S4, and S5.
- MP long-range plans (LRP) section.
- Communications section.
- SJA.
- IG.
- Public affairs.
- Unit ministry team (UMT).

6-5. The MP brigade (CS) command sergeant major (CSM) requires mobility and communications capabilities to execute his duties and responsibilities. The MP LRP section works with the corps G3 plans element, operating out of the corps main CP.

6-6. The MP battalion headquarters and headquarters detachment (HHD) provides C^2 for MP CS companies and any other assigned or attached MP elements. The battalion HHD consists of a detachment headquarters and a battalion headquarters that contains the battalion commander's staff. The staff officers supervise the major organizational elements, including the—

- S1, S2, S3, and S4.
- Communications and support sections.
- UMT.

6-7. As with the MP brigade (CS), the battalion CSM requires mobility and communications. The support section is vital to an MP battalion commander's ability to sustain his widely dispersed assigned or attached units during the performance of all five MP functions, primarily the MMS and AS functions.

6-8. The MP company (CS) provides support to an assigned AO. The company has a company headquarters, an MP operations center, a combat-medic section, and four platoons. The company headquarters provides maintenance, supply, communications, mess, and medical support to the unit. An MP operations center supports the unit's operation, conducting and planning for all five MP functions. The MP operations center includes three MP teams as the company-level response force.

6-9. The MP L&O detachments provide support to an assigned AO. The headquarters provides A/L support. The operation team plans and supervises desk operations, the traffic-accident and MP-investigation teams, and the force-protection teams. As with the EAC MP brigade (CS) detachments, requirements are based on the population supported and on METT-TC.

6-10. The number of assigned MWD teams is significantly less than those assigned to the EAC MP (CS) or (I/R) brigades. A kennel master, five explosive/patrol teams, and four narcotics/patrol teams are normally assigned to the corps MP brigade (CS). They are employed based on METT-TC.

COMMAND AND CONTROL

6-11. The C^2 in a corps MP brigade (CS) is consistent with that of any Army brigade. The MP brigade commander commands the brigade and all attached personnel. Battalion commanders work for the brigade commander. The MP company commanders receive their orders and work for their respective battalion commanders and direct their platoon leaders according to mission requirements.

6-12. When corps MP assets are sent to augment the division PM or the MP company, they are placed under the OPCON of the division PM for as long as the unit is needed in the division area. This command relationship is applicable to company-size organizations and smaller. If an MP battalion (or larger MP unit) augments the division, then the division PM will not exercise OPCON of that MP unit. The MP battalion will be placed OPCON as designated by the division commander but not under the division PM. One L&O detachment is normally attached to the supported division and placed OPCON to the division PM. However, MP leaders at each level establish an MP C^2 relationship based on METT-TC and the supported commander's needs.

EMPLOYMENT

6-13. Each corps MP brigade commander employs his assets according to METT-TC, the needs of the forces operating in his AO, and the priorities of the corps commander. Few MP assets in the corps area are employed to support fixed commitments. Instead, based on a broad consideration of the enemy and friendly situations, the corps MP are employed to support friendly forces engaged in combat, CS, and CSS operations. Since MP forces are dispersed throughout the corps area, the concentration of US forces, the location and vulnerability of critical sites, and the number of kilometers of the MSR to be controlled influence the designation of MP assets in the AO.

6-14. The MP brigade commander establishes his MP concept of operations based on the corps commander's concept of operations. His successful employment of MP assets depends on his foreseeing where the battle will be rather than where the battle is. Based on the rear-area IPB and PIAP, the MP brigade commander allocates and shifts resources to ensure the accomplishment of priority missions. This ensures the continuous support and forward sustainment of combat units and the safety of CSS units operating in the corps rear and sustainment areas.

6-15. To support the MP brigade commander's planning, the MP at brigade headquarters operate from several locations. Most of the staff locates in the vicinity of the corps rear CP where they can interface with the corps staff responsible for planning and executing rear operations. The control element of the brigade headquarters must be located where it can command and control its subordinate units. The brigade commander and his staff decide the best place to locate this element. The planning element of the headquarters locates near the corps main CP where it can interface with the corps commander's coordinating, special, and personal staffs. From there they monitor MP operations, integrate MP support with the corps plans cell for future operations, and learn the enemy situation through the G2's threat analysis almost immediately. The detailed information on rear-area activities and operations provided by the corps staff enhances the accuracy of the MP LRP.

6-16. The brigade S3 section provides the day-to-day planning and execution of an MP mission. The section provides a responsive CP that can relocate frequently throughout the AO. The S3 will normally provide liaison personnel to the corps rear CP, selected COSCOM units, interagencies, or other headquarters (based on mission requirements). When possible, battalion AOs coincide with the AOs of the CSG RAOCs. The MP brigade commander usually tailors battalions' AO boundaries to ensure responsive and flexible support across the corps's AO. He pays particular attention to the LOC behind the most heavily committed division and the critical bases and facilities in that area. He also ensures that the MP are available to respond quickly to combat operations throughout the entire corps rear area or during sustainment operations.

6-17. The MP brigade commander, coordinating with his battalion commanders, locates the MP companies where they can provide combat and CS power throughout the corps's AO. He bases his decision on the—

- Number and composition of urban areas.

- Location of CS and CSS units.
- Location of critical facilities (such as the headquarters, ammunition storage areas, and airfields).
- Expected threat.
- Level and frequency of support needed by the commander.
- Current and projected tactical situation.
- MSR network, including choke points and critical bridges and tunnels.
- Number of supported divisions and requirements.

6-18. The prioritization of MP missions is especially crucial during the early stages of the deployment when it is unlikely that an MP brigade commander will have a full complement of CS companies. Augmenting the division MP company with corps MP assets may not be possible at that time. Until it is, the MP brigade commander must locate corps MP assets to help meet the needs of the division while fulfilling the needs of the corps.

6-19. Like the brigade commander, battalion commanders plan the employment of their companies and platoons using METT-TC. Certain corps needs are constant. One MP company is allocated to provide security for the COSCOM, and one MP company is allocated to provide security for the corps's main CP. One platoon from that company may be used to secure the corps tactical CP or the jump CP. One or more platoons will also help secure the corps's rear CP. The number of MP assets allocated for a corps-level EPW/CI holding area and escort is based on METT-TC. However, a minimum of one platoon is dedicated to operate the corps's EPW/CI holding area and a minimum of one platoon per division is allocated for evacuating EPWs/CIs from division collection points. Additional MP assets may be allocated to provide security for the corps's ammunition storage area and ASPs supporting the divisions. The MP battalion commander places his companies where—

- MP assets support the brigade commander's concept of operations.
- The MP can support troop concentration, bases and base clusters, road networks, and critical areas.
- The MP can aggressively patrol critical terrain and monitor LZs and DZs to detect or deny enemy interference.
- The MP can respond to Level II threats.
- The MP can support the movement of combat resources throughout the AO.
- The MP can remove EPWs/CIs from division collection points.
- The MP can influence stragglers, refugees, and dislocated civilians.

6-20. Battalion commanders may choose to place a company behind the division rear boundary. This can help to ease the coordination between the corps MP and the division MP.

Chapter 7

Military Police Support to Division

Division MP provide essential support to the forward tactical commander. The MP provide a flexible, mobile, and lethal force ready to be employed in combat or CS operations.

USAMPS

The division is a large Army organization that trains and fights as a tactical team. Largely self-sustaining, it is capable of independent operations. The division is a unit of maneuver organized with varying numbers and types of combat, CS, and CSS units. Divisions fall mainly into two categories—heavy or light. The first category includes heavy divisions, their variants, and the Division XXI digitized heavy division. The second category includes the light infantry, airborne, and air-assault divisions. Each division can conduct operations over a wide range of environments. The success of Army operations depends on the success of its divisions.

OVERVIEW

7-1. In corps operations, divisions are normally comprised of three maneuver brigades, each with up to nine maneuver battalions, artillery battalions, aviation battalions, and supporting CS and CSS units. Divisions perform a wide range of tactical missions and are self-sustaining for limited periods of time. The corps augments divisions as the mission requires. All divisions must be able to deploy and conduct offensive, defensive, stability, and support operations. Airborne and air-assault divisions must be able to conduct forced-entry operations. Divisions may be part of a JTF or serve as the ARFOR headquarters.

7-2. MP support to a division is provided by an organic MP company that includes the PM and his staff (assigned to each division) and is augmented by the corps MP brigade. The division MP company provides support in the DRA and in the maneuver brigade rear area. It is fully mobile in order to relocate frequently under short notice. While division MP companies are capable of performing all five MP functions, they must be augmented by corps MP/CID to fully perform all five MP functions throughout the division's battle space. The division PM must receive at least one corps MP company for GS missions in the DRA, one L&O detachment for L&O missions, and one CID detachment for criminal investigations and LOGSEC. In a mature theater (or based on METT-TC), the division PM can expect two corps MP companies. In addition, the division PM may request a corps MP battalion headquarters to provide C^2 to corps MP units operating in the DRA or to perform missions as a TCF. Because the need for MP support exceeds division organic assets (and many times exceeds augmenting corps MP assets), careful planning of MP employment is essential.

DIVISION MILITARY POLICE COMPANY

7-3. The organization of the division MP company is set by specific tables of organization and equipment (TOE) and is designed to support a specific type of division. Regardless of the type of division, the assets in a division MP company enable a PM and his supporting staff to supervise the performance of all five MP functions. The PM coordinates the employment of the MP assets in the division's AO through his PM section. The division MP company contains a headquarters, a PM section, a medic section, and three to six platoons.

COMMAND AND CONTROL

7-4. The C^2 of the division MP company extends downward from the tactical commander to the division PM. The division PM is the division commander's special staff officer. Under normal circumstances, he works for the chief of staff. The division PM has OPCON of the division MP company and MP assets that have been provided by the corps. The division PM plans and employs all of the organic, assigned, or attached MP according to METT-TC. He anticipates the support needs of the division commander and plans the employment to meet his needs. After learning the division commander's concept of operations and estimating the situation, the division PM decides which MP functions are critical to accomplish the commander's mission. He then recommends the allocation of resources, tasks his assets, and establishes the priority of support.

7-5. When the division MP company is augmented by corps MP platoons, the platoons are placed under the direction of the division MP company commander, who assigns an AO to each platoon. When corps MP assets augmenting the division make up a company, the company is attached to the division. The company then comes under the OPCON of the PM, who assigns it an AO. The augmenting corps MP company is usually given an AO from the division rear boundary forward. However, this assignment is flexible and based on the division commander's needs as set forth by the division PM. The four platoons of that corps MP company perform all five MP functions in that AO. Additionally, the corps MP brigade (CS) L&O detachment supporting a division is attached to the division and OPCON to the division PM.

7-6. The division MP company commander directs the employment of company assets. Through his company headquarters, he provides administrative, maintenance, and logistical support to the PM section and to the company's platoons that are dispersed throughout the division's AO. These sections are mobile to support the platoons without delay despite the extended distances that may be separating them. The company commander is also mobile. He travels the extended distances from one platoon area to another when checking on the status of his soldiers and resolving the problems affecting mission accomplishment. The division HHC provides mess support. Personnel administration is handled by the company headquarters and the DISCOM. The number and kind of assets in a division MP company and the configuration of its platoons and squads are determined by the type of division to which the company is assigned.

NOTE: See the consolidated TOE update that is published yearly for complete details (base and objective) for all MP TOEs.

7-7. At the direction of the division commander, the division band may be available to perform its secondary mission of supporting MP operations. When the intensity of the conflict prevents the band from performing its musical mission, the division commander may direct it to augment the MP company. The band, normally employed as a unit under the direction of the bandmaster, is under the OPCON of the MP company commander while augmenting the MP. When tasked to augment MP assets, band members may be employed to augment MP security at the division main CP or to augment perimeter security at the division EPW/CI collection point. The band is released to perform its primary mission as soon as the tactical situation permits.

MILITARY POLICE EMPLOYMENT

7-8. The employment of division MP companies differs somewhat with the type of division to which they are assigned. In heavy divisions, the division MP companies provide GS to the DRA and DS to the division's maneuver brigades. When the platoon is providing DS to a maneuver brigade, the MP platoon leader coordinates all logistical support with the supported brigade. A platoon's DS relationship is seldom interrupted. However, METT-TC may require weighing MP support to the maneuver brigade tasked as the main effort or employing all platoons in a division MP company to function as a unit (such as during division-size river-crossing operations or deliberate breaching operations). In light divisions, the MP provide only GS to the DRA or during sustainment operations. Support exceptions are addressed under MP support to airborne and air-assault divisions, discussed later in this chapter.

7-9. In the division (where flexible support of an austere force is crucial), the division PM must have a clear understanding of situational awareness. To obtain current information for projecting MP needs in the division area, he must be mobile and be able to conduct split-cell operations. The assets available to the PM include the division MP company and at least one corps MP company. Corps augmentation is required for sustained operations and for special operations such as river crossings, dealing with dislocated civilians, and refugee internment or resettlement. The division PM coordinates with the corps PM and the MP brigade or CID commanders for—.

- Evacuating and guarding EPWs/CIs from division to corps.
- Providing law-enforcement assistance to HN forces in the division's AO.
- Providing corps augmentation for the division's AO, convoy security, LOC security, AS, R&S, L&O, and other missions. These other missions may include augmentation for security of the division main CP, ASPs, and other critical facilities.
- Integrating, sharing, and exchanging police intelligence between corps and division MP elements.
- Providing CID support.

SUPPORT TO THE HEAVY DIVISION

7-10. The Army's armored and mechanized infantry divisions (normally referred to as heavy divisions) provide mobile, armor-protected firepower that is normally employed for their mobility, survivability, lethality, and psychological effect (shock) on the enemy. These divisions destroy enemy armored and mechanized forces, and they can seize land areas and secure key terrain. Because of their strategic lift requirements, heavy divisions are slow to deploy from home staging bases into an AO. They have high consumption rates of supplies and have limited use in restrictive terrain. These capabilities and limitations are key factors in planning effective and efficient MP support throughout the division's AO.

7-11. In the heavy divisions, the PM section is organized to support split-cell operations at the main and rear CPs. The PM must be mobile to ensure that he is fully aware of the current status of critical MP operations. Therefore, the deputy PM (DPM) locates in one of the PM cells and—

- Handles the section's routine operations.
- Monitors ongoing division operations and MP support.
- Helps provide long-range planning and interface with the primary division staff.
- Forwards PM taskings to the division MP company commander.

7-12. The DPM and the operations sergeant normally set up operations at the division main CP where they can coordinate requirements with the division staff personnel. The company headquarters is initially located near the division rear CP in the division support area (DSA). Once augmentation arrives from the corps, the headquarters relocates with one of its GS platoons operating behind the brigade rear boundary. However, this location could be changed based on METT-TC, the supported commander's needs, and the company commander's idea of where he can exercise better C^2 for his unit.

7-13. The Army of Excellence (AOE) heavy division MP company has six platoons. Three platoons provide support to each maneuver brigade and are designated as DS. The other three platoons are designated as GS platoons. One MP platoon provides security for the division main CP; one provides security for the division's EPW central collection point; and one performs other MP operations within the division rear.

7-14. The GS MP platoons' AOs are configured based on METT-TC and the availability of MP augmentation from the corps. The DS MP platoons' AOs coincide with the supported maneuver brigade's boundary. Each platoon headquarters locates within its brigade's support area or any other area where it can best provide and receive support. To accomplish its mission, each DS platoon requires a minimum of two squads, each with three teams. One squad operates the EPW/CI collection point. The other squads perform MMS and AS operations. All MP platoons are capable of performing all five MP functions. However, performance of these functions is prioritized based on METT-TC and the division commander's concept of operations. The division PM, the company commander, and METT-TC dictate how these platoons should be tasked-organized to accomplish the mission.

SUPPORT TO THE LIGHT INFANTRY DIVISION

7-15. The AOE light infantry division (LID) is one of the most rapidly and strategically deployable divisions. It fights as part of a larger force in war or conducts missions as part of a joint force in MOOTW. Its C^2 structure readily accepts any augmentation forces, permitting task-organizing for any situation. The augmentation required for the division is largely determined by METT-TC. The division's capabilities allow it to exploit the advantages of restricted terrain and limited visibility. It achieves mass through the combined effects of synchronized small-unit operations and fires rather than through the physical concentration of forces on the battlefield. These characteristics are key factors in planning and employing MP assets in support of the LID.

7-16. The LID MP companies are capable of performing all five MP functions. However, their performance of these functions is prioritized based on METT-TC and the division commander's concept of operations. Contrary to the heavy division MP company, the LID MP companies are much smaller. The constrained size of the LID MP companies makes corps MP augmentation crucial to the sustainment of MP operations. Additionally, the LID MP company is the only MP unit with the capability of antiaircraft support through the use of shoulder-fired air-defense weapons.

7-17. The company has three GS platoons to support the division. No platoons are provided to the maneuver brigade. One platoon is normally located in the vicinity of the division main CP so that its resources can help support CP security. Another platoon locates in the DSA and operates the division EPW/CI collection point. The last platoon has an AO configured according to METT-TC and the commander's priority of MP missions. Each GS MP platoon has a headquarters and three squads, each with two teams. The PM section is located in the vicinity of the division main CP. The exact location is based on the current operational status and on METT-TC.

SUPPORT TO THE AIRBORNE DIVISION

7-18. The AOE airborne division can rapidly deploy anywhere in the world to seize and secure vital objectives. It conducts parachute assaults to capture initial lodgments, executes large-scale tactical raids, secures intermediate staging bases (ISBs) or forward operating bases (FOBs) for ground and air operations, and rescues US nationals besieged overseas. It can also serve as a strategic or theater reserve as well as a reinforcement for forward-presence forces. The airborne division can assault deep into the enemy's rear area to secure terrain or to interdict enemy supply and withdrawal routes. It can seize and repair airfields to provide a FOB and airheads for follow-on air-landed forces. It is capable of all other missions assigned to LIDs. The airborne division does not have sufficient armored protection to defeat heavier armored formations at close range. Therefore, engagements with enemy armored formations require special considerations. Antitank weapons in the division compensate for, but do not completely offset, this deficit.

7-19. Airborne divisions conduct operations in two phases—assault and defense. The division accomplishes the assault phase in three stages. First, they deploy; second, they establish the airhead; and third, they extend the logistics base and build their force.

7-20. Like other division MP companies, those assigned to the airborne division are employed to support their division commander's concept of operations. The airborne MP company has four platoons, each having three squads with two teams. Normally, the company headquarters and one of the platoon headquarters colocates with the PM section. The division PM's exact location depends on METT-TC. During the assault phase, the airborne division MP company provides DS to the assault brigade. The remaining platoons provide MMS and AS in the vicinity of the division main CP.

7-21. The nature of airborne operations makes the capture of EPWs likely. Therefore, during the first stage of the assault phase, the priority of MP support is given to EPW operations. After assembling the DZ or LZ, the MP collect EPWs captured during the assault. Combat elements are relieved of EPWs as far forward as possible. In airborne operations, EPWs are held for later movement to a central collection point. During the first stage of the assault, the MP perform limited straggler and refugee control and undertake AS operations, when possible.

7-22. MP support is reevaluated after the airhead is established in the second stage of the assault. The PM takes in consideration personnel and equipment flow, roadways, and security requirements to shift MP support priorities. As the entire operation matures, MP support may expand to include all five MP functions.

SUPPORT TO THE AIR-ASSAULT DIVISION

7-23. The air-assault division executes tactical missions at operational depths to achieve strategic results. It is capable of launching brigade-size air assaults of nearly 4,000 soldiers from either an ISB or a tactical assembly area (TAA). Within 6 hours, this air-assault task force (AATF) can attack an opposing force, occupy and defend key terrain, or establish a FOB (out to 150 kilometers) from which even deeper operations can be executed. Air-assault operations are terrain independent, but they rely on suitable weather and a detailed attrition/assessment of enemy capability—particularly air-defense assets along air corridors and in the objective area. The air assault is preceded by detailed, lethal and nonlethal condition settings, culminating in a comprehensive condition check before execution.

7-24. As the corps AATF unit, the brigade combat team (BCT) task-organizes and habitually trains with both aviation lift and attack (Apache) battalions. An air-assault division MP company has four platoons, each having three squads with two teams. To facilitate operations, air-assault division MP platoons are habitually aligned with each of the three AOE BCTs; however, as with aviation assets, the MP are task-organized to support the main effort's BCT/AATF. When conducting these operations, two MP platoons may be tasked to provide support; one is available to posture and participate in the air assault, and the other conducts AS and MMS operations at the pickup zone (PZ). This second MP platoon may further provide MMS and accompany a ground-assault convoy (GAC) to the objective, or it may remain at the PZ to receive EPWs returning on aircraft from the objective area. The mission of conducting MMS along this vulnerable ground LOC is particularly critical if the AATF objective is a FOB from which further division operations will expand.

7-25. MP participation in the actual air assault competes for heavy lift with FA and air-defense systems, as well as with vital Class V resupply. However, the lethality and versatility of the MP bode well for their employment, and two MP platoons are available to support the brigade air assault as the division's center of gravity.

7-26. When possible, habitually aligned platoons remain with their brigades, and corps assets perform GS missions. However, when no corps assets are available and two division platoons are employed as stated above, the two remaining platoons conduct division EPW collection-point operations and other MP functions based on METT-TC. Normally, the EPW platoon and the MP company headquarters colocate in the DSA. As required (and based on METT-TC), airflow planning includes EPW/CI evacuation from the AATF/ FOB collection point back to the DSA. The PM section operates from the division rear CP to facilitate I/R operations and to coordinate MMS and AS with key logistical staff. Due to potentially extreme distances on the air-assault battlefield, the DPM normally locates with the division main CP to serve as a key G3 battle-staff member and to coordinate PIO with the G2.

SUPPORT TO DIVISION XXI (HEAVY DIVISION)

7-27. Division XXI represents a significant change in the manner in which division operations are conducted. These changes are brought on by information-age capabilities; an increased integration of service components into an effective battle team; more lethal, survivable, and agile systems; and more capable soldiers and leaders. The Division XXI operates in a larger battle space and at a higher tempo than the AOE division. The division is evolutionary in design, but revolutionary in its use of information technology. It improves the Army's deployability while enhancing its ability to dominate in decisive fights. The following are characteristics of the Division XXI operational environment:

- **Multidimensional.** The division will operate in an extended battle space that goes beyond the traditional physical dimension of width, depth, and height. It includes portions of the electromagnetic spectrum; and it extends beyond the physical boundaries of the division through its communication and digital connectivity to other Army, joint, and coalition elements, even reaching back to CONUS.
- **Precise.** Precision operations go beyond a precision strike to include every aspect of military operations from deployment through combat and through redeployment or transition to other operations. Precision in decisive operations is enabled by three emerging capabilities. First, digitization provides soldiers and leaders at each echelon the information required for making decisions. Second, a full suite of strategic, operational, and tactical sensors linked to analytical teams fuses combat information into situational awareness across the battle space. Third, simulations enable Army elements to be tailored and operations to be planned, war-gamed, and rehearsed—yielding precision execution.

- **Nonlinear.** Nonlinear operations do not seek a battle-space grid of close, deep, and rear operations. Instead, the battle space is fluid, changing as METT-TC changes through the duration of mission preparation and execution. Another dimension of this characteristic is the synchronization of near-simultaneous operations to achieve nonlinear effects across the battle space.
- **Distributed.** Division operations are distributed or executed where and when required to achieve decisive effects concentrated at a decisive point. Dispersion empowers subordinates to operate independently within the commander's intent, leading to synergistic effects that exceed the effects of a centralized headquarters.
- **Simultaneous.** The concept of decentralization operations that are multidimensional, precise, distributed, and nonlinear yields the capability to conduct simultaneous operations across the battle space. Simultaneous operations seize the initiative and present the enemy leadership with multiple crises and no effective responses. Rather than a single, concentrated attack, the division executes a series of attacks (lethal and nonlethal) as simultaneously as possible.
- **Integrated.** Division operations are fully integrated with joint, multinational, and nongovernmental partners. Integrated operations enable the Army to leverage the full suite of capabilities that the services bring to the battle space.

7-28. The Division XXI heavy division is one of total integration—not rounding up or down. Reserve-component soldiers assigned to the division are an integral part of the division mission and task accomplishment. The division cannot operate without them. The division is focused on massing effects on the enemy—enabled by a new foundation of information technology and distributed logistics. The division is built around the pattern of "conduct decisive operations," accomplished by either fires or maneuver. The design performs complementary battle-space shaping in concert with the overall corps plan. Its long-range assets are not typically held in reserve nor just employed in the close fight; they are used to best support the corps and division commanders' plans of operations.

7-29. Although the basic structure of the division has not changed—three maneuver brigades, a division artillery, and a DISCOM—the internal structure and operational construction have changed in a near revolutionary way. The C^2 structure and the basic organizational structure are very similar to the divisions discussed above. However, by integrating the capabilities of digital technology and by using enhanced situational awareness, it is better able to execute combat operations.

7-30. The maneuver brigades have their own scouts—a brigade reconnaissance troop (BRT); therefore, they no longer rely exclusively on the battalion scouts. The maneuver battalions are reduced to three companies with a total number of 45 combat platforms. The division artillery has three Multiple-Launch Rocket System (MLRS) batteries per battalion. The cavalry squadron is under the aviation brigade's C^2 but retains and performs the division's traditional cavalry mission. Figure 7-1 depicts the organizational structure of Division XXI.

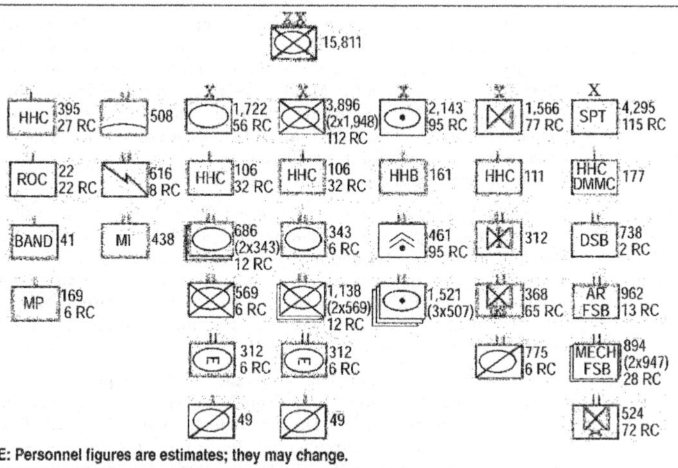

NOTE: Personnel figures are estimates; they may change.

Figure 7-1. Division XXI Structure

7-31. The division's CSS is centralized. Instead of each unit having its own organic CSS personnel, they are now centralized in CSS units. The CSS assets are reorganized and reassigned to the DISCOM. Forward-support companies (FSCs) are habitually associated with each maneuver battalion and forward-support battalions (FSBs) will support each maneuver brigade. This operational structure enables distribution-based rather than supply-based logistics concepts to support the division. The overall logistics footprint for the division is now reduced and the division has organic self-sufficiency to operate unsupported for up to 30 hours.

7-32. A major difference between the AOE fix-forward doctrine and the Division XXI maintenance support is the change in the actual repair site. Division XXI maintenance units directly supporting a given customer will diagnose faults and will replace components forward in the battlefield and will repair rearward. Units will be manned with maintenance personnel capable of performing DS and limited GS maintenance.

7-33. The division operates in a battle space two and a half times larger than an AOE division, and it will usually fight as part of a corps or a JTF. The division conducts offensive, defensive, stability, and support operations in cooperation with other elements of the corps/JTF to gain tactical or operational results significant to the joint-force commander's campaign goals.

7-34. Although the division is a heavy division, MP support is somewhat different from that of an AOE heavy division. The PM (based on METT-TC) establishes operations where MP assets can best be commanded. At the same time, overall MP operations can be synchronized to support operations within the division rear and the division commander's intent can be effectively supported. To this end, the PM conducts split-cell operations within the division main CP. Normally, the PM section is distributed between the security-operations cell (SOC) and the mobility/survivability (M/S) cell. The DPM (who works out of the M/S cell) is also part of the division planning team. He ensures that MP operations are integrated into division plans according to the PM's intent. The DPM attends coordination meetings and assumes coordination responsibility with the rest of the division staff when the PM relocates away from the division main CP.

7-35. The PM, as a staff officer, is involved with every aspect of planning and with the military decision-making process (MDMP). He uses the Force XXI Battle Command—Brigade and Below (FBCB2) C^2 system to understand the battlefield. He understands the mission and the commander's intent and priorities. The PM has complete situational awareness, establishes an effective staff relationship with the G3, and recommends the best way to employ MP resources.

7-36. The Division XXI heavy division MP company has five platoons. Although tasked-organized based on METT-TC and the commander's needs, three MP platoons provide DS for each maneuver brigade. The DS platoon requires a minimum of three squads (each having three teams) to accomplish the mission. The other MP platoons are normally employed forward in sector, providing AS and force protection for the division main and tactical (TAC) CPs and for other HVAs in the vicinity of the maneuver brigade rear boundary. As with the AOE heavy division, the Division XXI heavy division requires augmentation with at least one corps MP company for CS missions in the DRA and can expect two companies in a mature theater (based on METT-TC) as the main effort. In addition, the division may request an MP battalion headquarters to command and control CS MP operations in the rear or to conduct TCF operations.

7-37. While the division MP company is capable of performing all five MP functions, it must be augmented by the corps MP to fully perform all of these functions throughout a mature division's battle space. The PM prioritizes MP functions (based on the supported commander's needs), conducts a mission analysis, and requests needed corps assets through the division G3. Corps MP companies, L&O detachments, and CID assets are needed for sustainment operations. When corps MP companies and L&O detachments are attached to the division, they are placed under OPCON of the PM. The PM will then assign an AO and missions based on METT-TC.

7-38. The PM supports the division's distribution-based logistics concept by conducting aggressive R&S operations that secure the LOC. Since the main effort is to secure the LOC, traditional MMS missions (such as convoy security and escorts) are performed by exception. Through the allocation of increased MP to conduct LOC R&S operations aggressively, the division is able to detect and destroy threat forces before they interrupt operations in rear-area or sustainment operations.

7-39. Digitization gives the division the situational awareness that allows it to cover much larger distances and to move much faster. The systems that provide this digitization are generally unprotected HVAs on the battlefield, which require aggressive AS and force-protection operations.

7-40. The EPW operations are much more fluid, sometimes requiring the division GS MP to move forward and assume brigade collection points, thus allowing DS platoons to move forward with their brigades. Similarly, corps MP may be required to move forward and assume the division's central collection points as the division's rear boundary is pushed forward to shorten the LOC.

7-41. As with all AOE divisions, until corps L&O and CID detachments augment the division, L&O operations are on an exception basis. The MP collect intelligence during every facet of mission execution. The PM develops police intelligence in the division main CP through coordination with the SOC, the G2, the Assistant Chief of Staff, G5 (Civil Affairs) (G5), and the Assistant Chief of Staff, G6 (Communications) (G6).

Chapter 8

Separate Brigades and Initial/Interim Brigade Combat Teams

The MP support separate brigades and IBCTs designed to provide the brigade commander with MP assets that can serve as a force multiplier and that can be employed as an economy of force. The brigade MP platoon is as lethal, flexible, and capable as any other platoon in the MP Corps.

USAMPS

The Army uses separate brigades to inject a small but powerful force where it is needed. It must be able to fight and win while operating on its own for a sustained period of time. It must be able to defend itself on a 360-degree front in war or MOOTW. While there are some variations of separate brigades, it is the heavy separate brigade that is most commonly found within the force.

OVERVIEW

8-1. The Army's IBCT is a full-spectrum, wheeled combat force. It is employed in all operational environments against all projected future threats. However, it is designed and optimized primarily for employment in small-scale contingency operations in complex and urban terrain, confronting low-end and midrange threats that may employ both conventional and asymmetric capabilities. The IBCT deploys very rapidly, executes early entry, and conducts effective combat operations immediately on arrival to prevent, contain, stabilize, or resolve a conflict through shaping and decisive operations. The IBCT participates in war (with augmentation) as a subordinate maneuver component within a division or a corps and in a variety of possible roles. The IBCT also participates (with appropriate augmentation) in stability and support operations as an initial-entry force or as a guarantor to provide security for stability forces by means of its extensive capabilities.

SUPPORT TO SEPARATE BRIGADES

8-2. MP support to separate brigades is provided by a four-squad MP platoon organic to the brigade HHC. A separate PM cell within the brigade HHC serves as the C^2 element for the platoon (see Figure 8-1, page 8-2). Support to the platoon and the PM section for maintenance, supply, mess, and communications is provided by the brigade HHC. Since the platoon and PM section have no organic support, the MP leadership must perform close coordination for this support. However, the MP platoon must compete with other brigade HHC assets for priority of repair for weapons, vehicles, and communications equipment.

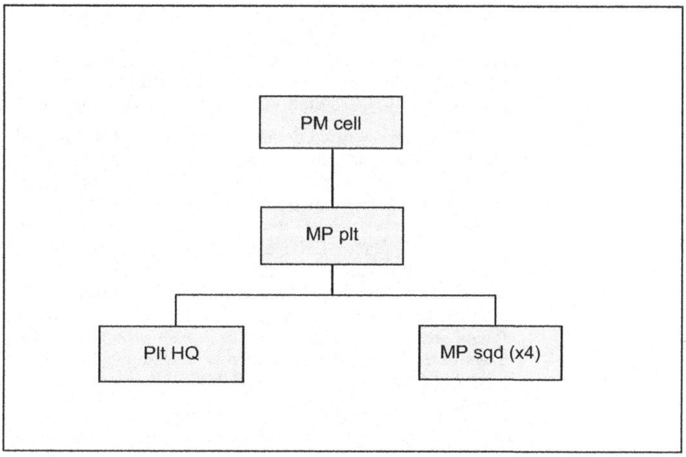

Figure 8-1. MP Support to a Separate Brigade

8-3. The C^2 of an MP unit supporting separate brigades also extends down from the tactical commander. The separate brigade PM has OPCON of brigade MP assets the same way the division PM has OPCON of division MP assets. This includes any MP assets that may have been provided from the corps. The PM advises the commander of a separate brigade on matters pertaining to MP operations. The platoon leader directs the execution of his platoon's mission based on the priorities set forth by the PM and the supported commander. Since the separate brigade's organic MP platoon is more robust than an MP platoon supporting a division maneuver brigade, METT-TC will determine the requirement for augmentation. However, corps L&O and CID augmentation is required.

8-4. The separate brigade's MP platoon is capable of performing all five MP functions. However, its resources are quite limited. Although the MP squads are employed according to METT-TC, the platoon supporting the separate brigade may have—

- One squad operating in the EPW collection point.
- One squad providing a mobile security screen and providing AS around the brigade's main CP.
- Two squads conducting MMS and AS throughout the brigade's rear area.

SUPPORT TO THE INITIAL/INTERIM BRIGADE COMBAT TEAM

8-5. The IBCT is a divisional brigade. It is designed to optimize its organizational effectiveness and to balance the traditional domains of lethality, mobility, and survivability with the domains required for responsiveness, deployability, sustainability, and a reduced in-theater footprint. Its two core qualities are its high mobility (strategic, operational, and tactical) and its ability to achieve decisive action through a dismounted infantry assault. The major fighting components are its motorized infantry battalions. The IBCT has a unique reconnaissance, surveillance, and target acquisition (RSTA) squadron to enhance situational understanding.

8-6. To achieve a rapid deployment threshold, the brigade's design capitalizes on the widespread use of common vehicular platforms—particularly a highly mobile, medium-weight, combat/CS platform coupled with the minimization of the personnel and logistical footprint on theater. Encompassing a personnel strength of about 3,500 and preconfigured in ready-to-fight combined-arms packages, the entire IBCT can deploy within 96 hours of "first aircraft wheels up" and begin operations immediately upon arrival at the APOD. In essence, the APOD is the TAA. The IBCT cannot conduct forced-entry operations, but it provides the JFC with an improved capability to arrive immediately behind forced-entry forces and begin operations to shape the battle space and to execute decisive action to expedite decisions. Once committed, the IBCT can sustain operations for up to 180 days without relief.

8-7. The IBCT's organization is expandable either through augmentation or scalability (according to METT-TC) in any given contingency. The IBCT is scalable in terms of its ability to accept like-type additional forces to expand core tasks and functions already resident within the IBCT (for example, adding additional infantry or RSTA organizations). The IBCT is also capable of accepting augmentation consisting of units or elements that execute tasks or functions not resident within the IBCT (for example, adding armor, air-defense, MP, or aviation assets). In both cases, added units execute their normal mission-essential task list (METL) tasks and, therefore, will not require extensive training to deploy and operate with the IBCT. In both cases, the IBCT includes the command, control, and communications (C^3) capabilities necessary to permit the rapid integration of additional enabling capabilities, particularly for operations outside the scope of small-scale contingency operations (such as stability operations, support operations, and war).

8-8. The IBCT organization excludes other unit-based capabilities often provided in a division slice. However, for each missing capability, the brigade headquarters includes staff cells capable of conducting rudimentary planning and analyses to ensure that all functional-area considerations are incorporated in route planning and preparation for operations. The first MP elements in support of the IBCT are at the brigade headquarters level.

8-9. The MP planning cell is composed of a two-person (a major [MAJ] and a sergeant first class [SFC]) planning cell located within the HHC's maneuver-support cell and under the direct staff supervision of the brigade S3. The role of the MP planning cell is significantly different from that of a traditional division PM or a separate brigade PM. The main difference is in the IBCT's lack of organic or habitual MP assets. However, the absence of organic MP assets makes the job of the MP planning cell that much more critical. The MP planning cell must—

- Understand the organization, capabilities, and limitations of the IBCT.

- Conduct effective liaison with higher headquarters PM elements.
- Become an effective planner and anticipator of MP requirements.
- Prepare MP staff estimates and employment recommendations.
- Plan for MP deployment via air, sea, rail, or land.
- Task-organize MP units effectively and efficiently.
- Assume C^2 of incoming MP forces or, if operating under a division or corps, relinquish C^2 to the division/corps PM or battalion commander (if appropriate for effective employment of MP forces).

8-10. Depending on METT-TC, the brigade may be augmented by MP elements ranging from a platoon to a battalion. In any case, the MP planning cell then becomes a staff planner and coordinator of functional matters pertaining to—

- Force protection, physical security, and vulnerability countermeasures.
- Operations for collecting, processing, and evacuating (possibly) EPWs, CIs, and detainees.
- US military prisoner operations.
- Operations for processing and controlling dislocated civilians and refugees.
- Customs and counterdrug operations.
- Criminal investigations and CID-/MP-investigator support.
- PIO and establishing links with HN police agencies and other international/interagency law-enforcement agencies.
- MMS, AS, and L&O operations.
- Coordination of MP or CID division/echelons above division (EAD) augmentation forces, MWD teams (explosives, narcotics, and patrol), L&O detachments, and MP I/R units through the division, corps, and major Army command (MACOM) PMs.
- Training guidance to other US forces required to perform MP functions in the absence of MP forces.

8-11. Since the IBCT is a divisional brigade, the division PM and the IBCT MP planning cell play an important role in developing an optimum MP force package to support the brigade commander's concept of operations. Despite the brigade's early-entry timelines, the MP planning cell must consider and plan for MP augmentation forces as early as possible to free valuable combat resources. Small-scale contingency operations that result in numerous EPWs, civilian detainees, and refugees will hamper momentum and freedom of maneuver.

8-12. Once the initial brigade receives MP augmentation (see Figure 8-2), the MP priority of effort during the offense will focus on ensuring that routes remain unencumbered and secure for movement of ground combat, CS, and CSS forces. The MP's priority of effort during the defense will focus on conducting AS and counterreconnaissance along the LOC, C^2 centers, and CSS bases. The MP may conduct response-force operations or become part of the TCF.

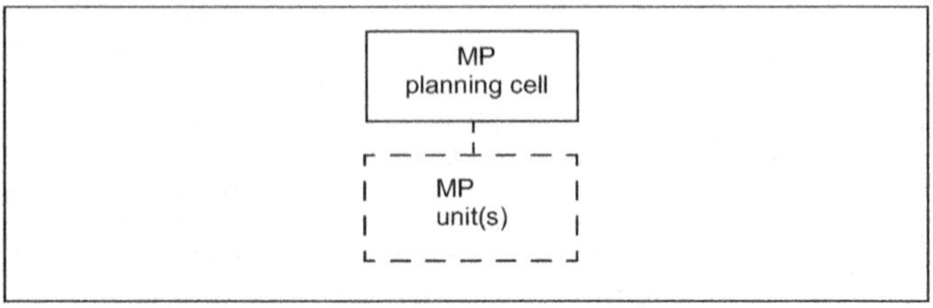

Figure 8-2. MP Support to the IBCT

8-13. Stability and support operations present some unique challenges. The IBCT may be deployed to a geographical area that is politically unstable, that lacks civil control, or that is in complete turmoil. The MP planning cell plans for and coordinates MP support according to available resources and the supported commander's needs. In this scenario (and based on METT-TC), an MP battalion TF could be required to deal with the challenges of—

- A significant number of refugees or dislocated civilians.
- AS or force-protection issues.
- MMS operations.
- Black-market and criminal investigations.
- Restoration of order.
- Civil unrest.
- Intervention of private and nongovernmental organizations.

Chapter 9

The United States Army Criminal Investigation Command

The history of the USACIDC goes back to World War I, when General John J. Pershing organized the CID in France. Today, the USACIDC is the Army's sole agent responsible for investigating felony crimes on and off the battlefield. The USACIDC provides investigative support to commanders at all echelons.

USACIDC

The USACIDC investigates offenses committed against US forces or property, or those committed by military personnel or civilians serving with US forces or where there is a military interest. The USACIDC agents investigate violations of international agreements on land warfare. They conduct special investigations at the direction of the USACIDC's commanding general (CG) or a higher authority. In general, the USACIDC's missions include—

- Investigating and deterring serious crimes.
- Conducting sensitive/serious investigations.
- Collecting, analyzing, processing, and disseminating criminal intelligence (CRIMINTEL).
- Conducting protective-service operations for designated personnel.
- Providing forensic-laboratory support.
- Maintaining Army criminal records.
- Enhancing the commander's crime-prevention and force-protection programs.
- Performing LOGSEC operations.

OVERVIEW

9-1. The USACIDC's operations help the commander maintain discipline and order by preventing or investigating crimes that reduce a unit's ability to fight. During the investigation of serious crimes, the USACIDC concentrates its efforts on investigating serious crimes such as wrongful deaths, controlled-substance offenses, theft, fraud, sex crimes, and assaults. The USACIDC also conducts sensitive and special investigations involving matters pertaining to senior Army officials and those associated with classified programs.

9-2. The USACIDC agents collect, analyze, process, and disseminate criminal intelligence/information relating to crime within or directed toward the Army. Specific information relating to modus operandi, crime techniques, investigative leads, gang violence, and terrorism is shared with the appropriate intelligence and law-enforcement agencies. Conversely, USACIDC agents solicit and receive crime-related information from the MP and from local, national, and foreign law-enforcement and intelligence agencies. Special agents identify and evaluate crime-conducive conditions and indicators of potential attacks against Army property, facilities, or personnel. They then provide reports to the appropriate commander.

COMMAND AND CONTROL

9-3. The USACIDC is a centralized (stovepipe) MACOM whose special agents in the field report through the USACIDC's chain of command (detachment to battalion to group) to the CG, who in turn reports directly to the Army Chief of Staff and the Secretary of the Army (see Figure 9-1).

9-4. As an independent MACOM, the USACIDC's tactical units are not under the C^2 of supported organizations. The USACIDC's elements may be temporarily attached to a supported commander when required to accomplish a nonstandard mission. Reassignment, promotion, accreditation, and disciplinary actions are retained by the USACIDC. Attachments will be coordinated with the appropriate USACIDC headquarters planning the specific mission and approved by the CG, USACIDC. Additionally, although there is no formal staff relationship, USACIDC commanders advise their supported commanders on criminal-investigation matters. This enhances the quality, reliability of information, support, and trust between USACIDC elements and their supported commanders.

WARTIME SUPPORT

9-5. The USACIDC supports each echelon of command from the division to the ASCC. The theater USACIDC structure is comprised of a C^2 headquarters

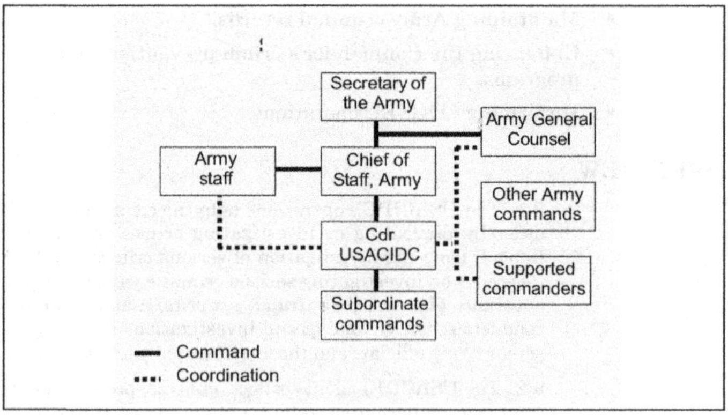

Figure 9-1. USACIDC Chain of Command

and mobile, modular, and tailorable investigative detachments. The USACIDC supports combatant commanders with the following functions:

- **LOGSEC.** Tracking and protecting materials and equipment from the manufacturer to the soldier on the battlefield.
- **CRIMINTEL.** Collecting, consolidating, analyzing, and disseminating intelligence associated with criminal and terrorists activities targeted at Army interests.
- **Criminal investigations.** Investigating war crimes and, in some cases, crimes against coalition forces and HN personnel.
- **Protective-service operations.** As in peacetime, protecting key personnel anywhere on the battlefield.

9-6. The USACIDC's LOGSEC function protects the Army's supply pipeline against criminal activities from the manufacturer, through logistics channels, all the way to the frontline soldier. It involves preventing, detecting, and investigating criminal and terrorist activities such as supply diversion, destruction, and sabotage or product substitution. The USACIDC's LOGSEC operations assess LOGSEC, identify weaknesses, and provide a prioritization of threats so that commanders can implement preventive measures to reduce the vulnerability of the logistics pipeline. Whenever possible, the USACIDC will initiate actions to recover logistical losses and return them to Army control.

The Army let a $22 million contract for the manufacture of its mobile kitchen trailer (MKT)-90 mobile field kitchens in 1990. These kitchens were distributed to Army units worldwide, including those units serving in Saudi Arabia during Operation Desert Shield/Desert Storm. After distribution of the MKT was well underway, a US government quality-assurance representative received information from the manufacturer's employees indicating that the kitchens were not built to contract specifications. This information led to an investigation by the USACIDC's major fraud procurement unit (MFPU). The investigation revealed that inferior material and manufacturing processes were substituted, which allowed the contractor to realize an additional $228,000 in profit. The US Army Environmental Hygiene Agency determined that the MKT-90 had serious sanitation defects and posed a safety hazard to troops using and being fed from these kitchens. The problem with the MKT was identified and expeditiously resolved to ensure the safety of soldiers in the field. The primary subject was indicted by a grand jury and subsequently pled guilty to one count of major fraud. He was found guilty, sentenced to 10 months confinement, and ordered to pay restitution.

9-7. The USACIDC's CRIMINTEL collection efforts focus on the identification and prevention of terrorist and nonterrorist crimes against US and allied military personnel, facilities, and other interests. Upon collecting this information, agents recommend countermeasures to combat subversive activities through coordination with MI, the PM, rear-area operations officers, and HN military and civil intelligence agencies. Additionally, when directed, the USACIDC becomes the lead US military investigative agency at theater level. When this happens, it is tasked with leading the prevention-of-terrorism effort from all services, not just the Army component.

> The MI analysts, the MP, and US Army criminal investigators deployed in Kosovo formed a combined all-source information center. This center provided threat assessments to units assigned to TF Falcon. Fifty-five percent of these assessments referenced some form of criminal activity. These intelligence products provided commanders with reliable information concerning potential incidents or criminal activity. With nearly 7,000 US soldiers deployed within the region, these assessments were absolutely crucial for the force-protection planning effort.
>
> The fusion cell within the information center developed intelligence products from national, theater, and operational sources. Due to the significant threat posed by criminal elements in the region, two CID military agents and two CID civilian analysts were attached to the fusion cell to facilitate the police-intelligence function. The CID personnel, in cooperation with MP soldiers, played a key role by linking CRIMINTEL to specific groups and events. The CRIMINTEL collection effort was specifically targeted on weapons, drugs, organized crime, and the identification of smuggling routes. The identification of smuggling routes resulted in a significant increase in the number of weapons being confiscated. The timely transfer of CRIMINTEL products to tactical units enabled a rapid response to serious confrontations, increased confiscation of arms and ammunition, and improved stability in TF Falcon's AO.

9-8. During war and MOOTW, the criminal-investigation effort includes—

- War crimes and, when directed, crimes against coalition forces and HN personnel.
- Major felony crimes committed by US personnel or EPWs/CIs and, when directed, crimes committed by foreign nationals in which there is an Army interest.
- Criminal acts by indigenous personnel, factions, and ad hoc groups.
- Special/sensitive investigations, hostage negotiations, polygraph operations, and force-protection operations.
- CRIMINTEL operations.

9-9. The USACIDC detachments provide enhanced coverage and protection for designated key and essential leaders during war and MOOTW. The protection requirement for senior JTFs and Army commanders may be significantly greater during MOOTW than during peacetime or war as the propensity of asymmetrical threats (such as criminal and terrorist groups) operating in the AO also increases.

ORGANIZATION

9-10. The USACIDC is organized with an array of multifunctional units. As with many other Army organizations, its force is in a transition process. During peacetime, the USACIDC groups provide brigade-level area support for worldwide Army operations. The group provides C^2, staff planning, and administrative support for USACIDC elements assigned to an AOR or deployed to a contingency operation. During war, a group provides the same capabilities in support of an ASCC and provides C^2 for all USACIDC units in the theater. Regardless of the operational environment, the group ensures the connectivity between all USACIDC units. It establishes and maintains links with supported units and interagency, joint, multinational, allied, and HN authorities on matters pertaining to Army and USACIDC operations. The group headquarters has a command section, a detachment headquarters, an S1, an S2, an S3, an S4, an SJA, and a communications section. It provides C^2 for up to six subordinate battalions. Major differences in the AOE and Force XXI groups are in an increase of support and mission personnel and the movement of all polygraph support down to battalion level.

9-11. During peacetime, the USACIDC battalions provide area support for worldwide Army operations. They provide C^2, staff planning, forensic-science support, CRIMINTEL, polygraph support, technical guidance and supervision of USACIDC operations, and administrative support for USACIDC elements assigned to an AOR or deployed to a contingency operation. During war, the USACIDC battalion provides the same capabilities in support of each corps and TSC. Under AOE designs, one battalion supports each corps and TSC. Under the Force XXI design, up to two battalions support each TSC. Like the groups, each USACIDC battalion ensures connectivity and establishes and maintains links with other units/agencies within its AOR. The battalion headquarters consists of a command section, a detachment headquarters, an S1, an S2, an S3, an S4, and a communications section. It provides C^2 to five USACIDC detachments on a sustained basis. Its span of control can accommodate up to seven detachments for a short period of time. Major differences in the new battalion are the addition of a CSM, an S6 officer, a detachment commander and support personnel, and CRIMINTEL and forensic personnel.

9-12. The USACIDC investigative team is the smallest operational investigative element. Each team consists of two special agents (a warrant officer and a noncommissioned officer [NCO]). METT-TC requirements dictate that these teams have the capability to operate independently from the detachment headquarters. The investigative teams are the building block for both AOE and Force XXI USACIDC detachments. Both designs provide levels of flexibility to task-organize without splitting units apart. The detachment (with its complement of teams) performs the full range of criminal-investigative functions in all operational environments, but it has limited CRIMINTEL and LOGSEC management capabilities and no forensic or polygraph capability. Each detachment includes a headquarters section and up to eight investigative teams. METT-TC requirements dictate that these teams have the capability to operate independently from the detachment headquarters. The major differences between the AOE and Force XXI detachments' design begin with replacing the heavy and light division support elements (DSEs), the corps-area support element (CASE), the TSC-area support element, (TASE), and the port-area support element (PASE) designs with a modular design consisting of a standardized headquarters element and two special-agent sections. Each section consists of four investigative teams. Personnel administration, supply, and the bulk of administrative support found in the AOE organizations have been consolidated at battalion.

Additionally, the Force XXI detachment is commanded by a special-agent warrant officer.

9-13. Both the AOE and the Force XXI units retain the Army's standard dependencies upon other units for support requirements (religious, health services, finance, photographic processing, and so forth). The new USACIDC TOEs have enhanced A/L support capabilities. However, the new units still retain the AOE support requirement for food services and emergency Class III/V resupply from CS MP brigades, battalions, and companies.

THE FIELD INVESTIGATIVE UNIT

9-14. Sensitive investigations are normally conducted by the field investigative unit (FIU). The FIU is a one-of-a-kind organization within the DOD that enables the Secretary of the Army to conduct sensitive investigations requiring access to special information or programs that are highly classified. The FIU works closely with the Army IG, the Judge Advocate General of the Army, and the Army General Counsel to support commanders in the special-operations and intelligence communities and in the area of classified acquisition programs. The FIU may also be assigned investigations involving senior Army personnel or those of special interest to the Army leadership.

THE COMPUTER-CRIME INVESTIGATION UNIT

9-15. The computer-crime investigative unit (CCIU) deals with intrusions involving classified networks or multijurisdictional offenses. The CCIU works closely with MI and federal law-enforcement agencies to coordinate military actions, nonmilitary-affiliated offenders, and foreign-intelligence services. In addition, it provides technical assistance to CID elements that are conducting computer-related investigations.

THE PROTECTIVE-SERVICE UNIT

9-16. The protective-service unit (PSU) provides worldwide protective services to designated personnel to protect them from assassination, kidnapping, injury, or embarrassment. The PSU plans, coordinates, and executes executive protection for—

- The Secretary of Defense.
- The Deputy Secretary of Defense.
- The Chairman of the Joint Chiefs of Staff.
- The Vice Chairman of the Joint Chiefs of Staff.
- The Secretary of the Army.
- The Army Chief of Staff.
- The Army Vice Chief of Staff.
- Others (as directed).

9-17. During war and MOOTW, the USACIDC routinely provides protective-service support to senior Army ground commanders and, in some cases, to JTF commanders. The USACIDC agents also provide training to the MP details providing protective services to senior field commanders.

The President of the US hosted the North Atlantic Treaty Organization's (NATO's) 50th summit anniversary celebration in Washington, DC during the period of 22 to 26 April 1999. The President invited 28 heads of state or prime ministers, 48 foreign ministers, 88 ministers and chiefs of defense, and the chairman of the military committee to the summit. During the conference, the principal guests resided in 23 hotels in Washington, DC. The responsibility for providing protection for the ministers, chiefs of defense, and the chairman for the military committee was given to the USACIDC's PSU. In preparation for this mission, additional training on protective-service operations and evasive driving techniques was given to over 200 reserve agents. These individual mobilization augmentees, troop program units, and other USACIDC agents were critical to the successful execution of this mission.

US ARMY CRIMINAL-INVESTIGATIONS LABORATORY

9-18. Forensic-laboratory support is provided by the US Army Criminal-Investigations Laboratory (USACIL) that is currently located at Fort Gillem, Georgia. The USACIL is a nationally accredited forensic laboratory that provides all three services with dedicated, highly responsive, deployable, state-of-the-art scientific and forensic investigative support. The six forensic divisions of the USACIL include the following:

- Chemistry (serology, drugs, trace, and deoxyribonucleic-acid [DNA] testing).
- Questioned documents (alterations and comparisons of handwriting, typing, and printing).
- Firearms and tool marks (weapons and obliterated numbers).
- Photography (evidence, charts and photos for court).
- Latent prints.
- Computer forensics.

9-19. The USACIL collects, receives, and examines evidence and prepares reports of findings. It also provides expert-witness testimony at court-martials.

US CRIME RECORDS CENTER

9-20. The US Crime Records Center (CRC) maintains the Army's criminal records. It maintains more than two million MP reports and USACIDC reports of investigations. These records are retained for at least 40 years before being destroyed. During their retention, the records are actively maintained and frequently researched.

During the course of an investigative search, civilian police authorities seized 19 hand grenades. A routine search of stolen property and criminal index files failed to identify the grenades as having been stolen from a nearby military installation. However, a search of the CRC files identified a USACIDC investigation relating to the theft of a case of grenades that occurred 20 years earlier. Further investigation revealed that the lot number of the recovered grenades was identical to that of the stolen grenades. The investigative leads resulting from the CRC file search contributed to the resolution of both cases.

9-21. The CRC conducts more than 10,000 criminal-history name checks each month to identify victims and perpetrators of criminal offenses. The checks are requested not only by USACIDC agents, but also by other military and civilian law-enforcement officials. The CRC is also the Army's agent for Freedom of Information Act (FOIA) and Privacy Act requests relating to criminal investigations and MP reports. The CRC also manages the Army's polygraph programs and their support to Army installations around the world.

EMPLOYMENT

9-22. Whenever possible, USACIDC units are colocated with the MP. This provides unified MP support across the battlefield, and it facilitates logistical support (primarily mess) provided by MP units. However, the USACIDC maintains its vertical C^2. Based on METT-TC, the MP group (CID) will colocate with the supported ASCC and the MP battalions (CID) supporting the corps and EAC will colocate with the MP brigade (CS). When feasible, MP detachments (CID) will colocate with MP battalions/companies and division MP companies. Figure 9-2 illustrates a typical battlefield array for AOE organizations, and Figure 9-3 illustrates a Force XXI array.

9-23. Since USACIDC detachments are austere organizations with little organic A/L capability, they rely on the battalion for unit-level maintenance, supply, and personnel support. However, detachments must often operate independently when performing criminal-investigation functions throughout the AOR. During these times of long-distance separation, detachments must seek A/L support from the supported unit.

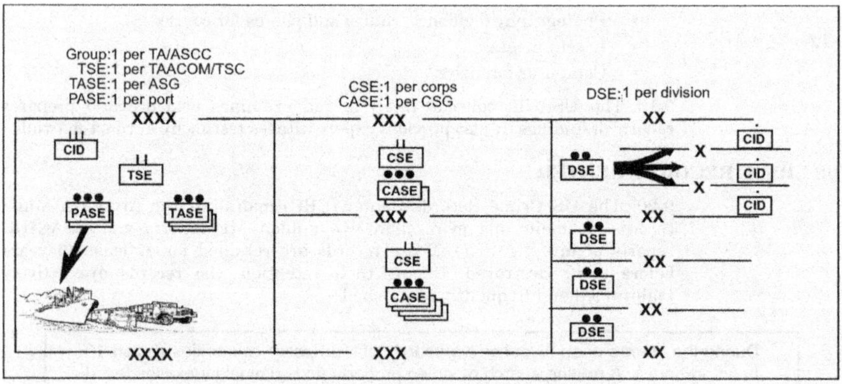

Figure 9-2. AOE Battlefield Array

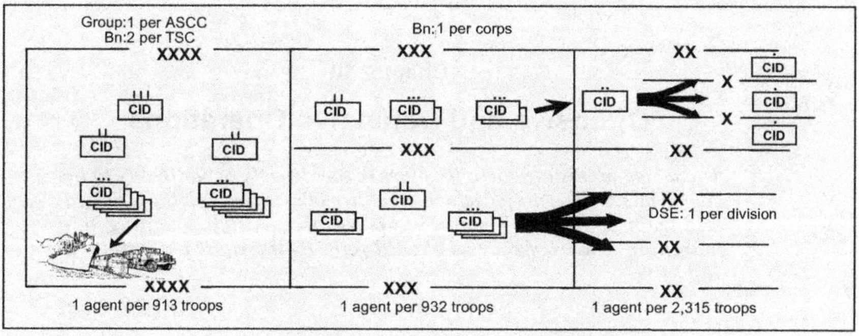

Figure 9-3. Force XXI Battlefield Array

Chapter 10

Offensive and Defensive Operations

The whole of military activity must relate directly or indirectly to the engagement. The end for which a soldier is recruited, clothed, armed, and trained; the whole object of his sleeping, eating, drinking, and marching is simply that he should fight at the right place and the right time.

Clausewitz

Army doctrine considers the full range of operations from war to MOOTW. An operation is a military action or the carrying out of a military mission needed to gain the objectives of any battle or campaign. Offensive and defensive operations normally dominate military operations in war and in some smaller-scale contingencies. An offensive operation is aimed at destroying or defeating the enemy and imposing our will on him for a decisive victory. Defensive operations defeat an enemy attack, buy time, economize forces, or develop conditions favorable for a counteroffensive that regains the initiative and defeats the enemy.

OVERVIEW

10-1. The offense is the decisive form of battle. Offensive campaigns and major operations are designed to achieve operational and strategic objectives quickly and decisively at the least cost. Offensive operations combine the terrain and the force to achieve the objective. The main purpose of the offense is to defeat, destroy, or neutralize the enemy force. Offensive operations are characterized by a rapid shift in the main effort to take advantage of opportunities by momentum and simultaneous destruction of enemy defenses.

10-2. The immediate purpose of defensive operations is to defeat an attack. Military forces defend only until they can gain sufficient strength to attack. Commanders choose to defend when they need to buy time, to hold a piece of terrain, to facilitate other operations, or to erode enemy resources at a rapid rate while reinforcing friendly operations. The ARFOR conduct defensive operations as part of major operations and campaigns and in combination with offensive operations (see FM 3-0).

SUPPORT TO OFFENSIVE OPERATIONS

10-3. MP support to offensive operations varies according to the type of operation being conducted. The MP functions support each echelon commander based on METT-TC. In the offense, MP priorities are often placed on MMS, I/R, and AS operations. Regardless of the offensive effort, MP leaders supporting an offensive operation must—

- Understand the commander's intent and his concept of operations.
- Anticipate that selective elements of the offensive force may need to pause, defend, resupply, or reconstitute while other forces attack.
- Anticipate changes in the operational tempo and prepare to weight the MP effort to meet the maneuver commander's intent.
- Provide MP support within the context of linear and nonlinear operations. They must understand how both types of operations affect the security of rear areas and LOC, the protection of C4I, the means of sustainment, and the land force's mobility.
- Support the maneuver commander's intelligence efforts by conducting the PIO function. The MP must observe, identify, and report information.
- Assist maneuver forces to mass. The MP protect C^2 nodes such as the CP that orchestrates the mass efforts and resources that sustain the offensive move. The MP must deny incursions by enemy reconnaissance forces into the AO. They also maintain surveillance, provide early warning, and impede and harass the enemy with supporting and organic fires.
- Speed the commitment of follow-on forces and reserves. They must also conduct MMS operations to assist the commander in a smooth and quick shift of forces to support the main effort.
- Anticipate the pursuit and exploitation by positioning MP forces that can support continuous operations. The MP support as far forward as possible while protecting resources needed to seize the opportunity. The EPWs/CIs are collected as far forward as possible. The MP coordinate the treatment of dislocated civilians and refugees with the HN or foreign forces.
- Tailor the MP support based on the unit being supported (corps, divisions, and brigades).
- Consider the location and composition of response forces and the TCF and the strength of rear-area threats.
- Integrate a force into the offensive plan that is capable of conducting aggressive combat operations against designated targets.

> During Operation Just Cause, an MP platoon received the mission to disable an armed Panamanian Defense Force (PDF) patrol boat docked near a key objective. After receiving the mission, the platoon leader maneuvered his force into position and attacked after receiving enemy fire. The platoon successfully disabled the patrol boat and captured the objective while suffering minimum casualties.

10-4. During offensive operations, MP support provided by teams, squads, and platoons provide the commander with an agile, flexible, versatile, and capable force ready to contribute to the overall mission success.

SUPPORT TO DEFENSIVE OPERATIONS

10-5. The MP support defensive operations consistent with the commander's intent and priorities. They support a defending force by assisting subordinate elements in the execution of different, yet complementary actions. For example, the MP may support the maneuver of an attacking unit by conducting MMS operations while simultaneously providing AS for a defending unit in a base cluster. The MP support reflects the maneuver commander's focus—destroying an attacking force, retaining or denying key terrain, moving away from an enemy force, or a combination of the aforementioned. However, despite the intent of the defensive operation being conducted, MP leaders must—

- Conduct MMS operations to help a force maneuver and mass. The MP must anticipate the shift from the defense to the offense and assist the forward movement of reserves.
- Conduct AS operations to deny information to enemy reconnaissance elements seeking out the exact location of the defending force. The MP are positioned where they can control key terrain or improve the defensive capability of bases and base clusters. The MP conduct aggressive R&S to locate and deny enemy access to critical logistical facilities.
- Protect sustainment resources while supporting the lateral, forward, and rearward movement of forces.
- Consider the type and size of the AO, the LOC security, the rear threat, and the number of EPWs/CIs and dislocated civilians to determine how they may affect the movement of forces.

> During Operation Desert Shield, EAC, corps, and division MP units conducted a myriad of missions to support both the XVIII Corps's (Airborne) and the VII Corps's movement to their TAA. The MP's main focus was on MMS and AS throughout the AO to ensure freedom of movement and to protect US and coalition forces against terrorists and other threats in the rear area.

10-6. The success of MP operations in the defense depends on the MP leader's situational awareness, the commander's intent, and the precise employment of MP resources when and where they are needed. In the defense, the MP perform their five functions to assist the defending commander by providing a lethal mobile force that allows him to concentrate his efforts and resources in fighting the next battle.

Chapter 11

Stability and Support Operations

ARFOR conduct stability operations in a complex, dynamic, and often asymmetric environment. Stability operations are usually nonlinear and noncontiguous. They are often time and manpower intensive. The purpose of support operations is to meet the immediate needs of designated groups for a limited time until civil authorities can accomplish these tasks without Army assistance.

FM 3-0

The Army's mission is to prepare for war and, when deterrence fails, to achieve a quick, decisive victory. The DOD projects that many of its future missions will be stability and support operations. However, these operations are not new to the Army. The Army has participated in stability and support operations in support of national interests throughout history. The Army has protected its citizens, assisted nations abroad, and served America in a variety of other missions. The pace, frequency, and variety of stability and support operations have quickened in the last three decades. As demonstrated in Operations Provide Comfort, Restore Democracy, and JTF Los Angeles, these types of operations present a challenge to all services.

OVERVIEW

11-1. Stability operations are conducted outside of the US and its territories to promote and protect US national interests by influencing political, civil, and military environments and by disrupting specific illegal activities. Stability operations may include both developmental and coercive actions. Developmental actions enhance a HN government's willingness and ability to care for its people. Coercive actions apply carefully prescribed limited force or the threat of forces to change the AO's environment.

11-2. Support operations provide essential supplies and services to relieve suffering and to help civil authorities prepare or respond to crises. In most cases, the ARFOR's efforts are focused on overcoming conditions created by man-made or natural disasters. The ultimate goal of support operations is to meet the immediate needs of designated groups (for a limited time) until civil authorities can accomplish these tasks without Army assistance.

11-3. Stability and support operations are distinctly different from offensive and defensive operations and must be analyzed differently. For instance, the application of the rules of engagement (ROE) will be unique to the operation. The restrictions on combat operations and the use of force must be understood and obeyed at all levels. MP leaders must clearly take the initiative and ensure that the applicable guidance is implemented and followed by all subordinates.

11-4. Stability and support operations have more diverse political considerations than are usually encountered in war. There are also complexities that must be addressed in joint, combined, and interagency stability and support operations. The Army's challenge is to execute these operations successfully in a joint or combined arena without degrading its capability to wage war successfully.

11-5. If the US is to fight and win future wars, it must develop, train, and employ its force structure in a manner that ensures success in every operation across the spectrum of military operations. Under these auspices, MP forces can provide commanders with especially valuable assets for stability and support operations. The MP have the unique capability of serving as a combat, CS, and CSS force. The domestic and international acceptability of the MP operational image frequently makes the MP the most appropriate force for stability and support operations. Additionally, the MP are flexible and capable of rapidly transitioning from one end of the spectrum to the other if the environment changes.

FORCE SUITABILITY

11-6. The MP's capabilities and their operational and supporting tasks are as integral to stability and support operations as they are to offensive and defensive operations. MP branch-/core-related skills are highly compatible with the capabilities required for stability and support operations. The MP provide a highly capable, politically acceptable force that is suitable for a variety of missions. They possess robust moving, shooting, and communicating capabilities; and they project an assist, protect, and defend image that is particularly important when tailoring a force that requires significant capabilities but a low political profile. The MP provide a flexible, but limited, economy-of-force organization. Mission requirements will always exceed available MP resources, especially in stability and support operations. For this reason, MP assets must be prioritized to deliver the greatest mission capability. The specific operations that MP units perform at a given time are determined by the echelon commander's needs, the mission requirements, and the availability of MP resources. As discussed in previous chapters, the echelon commander, through the command's PM, sets the priorities for MP operations. As in offensive and defensive operations, the PM knows that while the MP force can perform all MP functions, they may not be robust enough to perform all assigned MP missions all of the time. The PM must continuously make trade-offs between the numbers and kinds of MP missions that can be supported.

11-7. The MP's capabilities are further enhanced by their training, mind-set, and experience in dealing with people in highly stressful and confusing situations. Although capable of conducting combat operations when needed, the MP are highly practiced in de-escalation and in employing the minimum-essential force to contain potentially violent situations. This mind-set serves as the framework for MP law-enforcement training and is especially applicable in MP support for stability and support operations. Furthermore, this mind-set is exercised and reinforced daily in peacetime law-enforcement operations. This is a significant capability. MP soldiers learn and receive constant reinforcement training in controlling a situation on the spot before it escalates and cannot be controlled. The MP have a unique understanding of human nature and are adept at handling emotion-laden situations.

> During operations in Kosovo, as both Serbs and Albanians attempted to create unfavorable conditions and responses by US forces, the MP kept their cool under fire in the streets of Gnjilane and Kamenica. The MP's reaction led to a quick understanding by all sides that the US Army MP were fair and just and understood the tense situations.

11-8. The MP stress four essential training competencies—skill proficiency, human dimension and attitude, camaraderie and teamwork, and leadership. These competencies are the foundation for all MP operational performances. They equip the MP to exercise discretion in dealing with others, to protect and assist those in need, and to accomplish the varying demands of MP missions in stability and support operations and, when needed, in war.

STABILITY AND SUPPORT

11-9. Stability and support operations are performed in support of US authorities or in support of international bodies and foreign governments when so required by US authorities. The MP have long provided the essential support for stability and support operations; however, the frequency of these operations has increased significantly in recent years. Consequently, the MP have been deployed in greater numbers with greater frequency. MP units have a substantial history of successful, simultaneous, and continuous deployments in both a lead and a support role.

SUPPORT TO DOMESTIC CIVIL AUTHORITY

11-10. The purpose of support to domestic civil authorities is to—

- Render humanitarian assistance.
- Provide disaster relief.
- Restore order.
- Combat terrorism.
- Support counterdrug operations.

11-11. Military operations supporting domestic civil authorities are governed by the provisions of the Stafford Act, the Posse Comitatus Act, and other laws and regulations. Peacetime experience in civil-military affairs equips the MP to oversee, assist, and train other units and agencies for law enforcement, sensitive security, and operations to control dislocated civilians (within the legal limits of a particular stability or support operation). For this reason, the MP are often needed after other forces have redeployed, and they stay until government order is restored. Since active-component military missions are constrained by the Posse Comitatus Act, nonfederalized national guard (NG) units are often employed to conduct these operations (particularly counterdrug operations). The NG units can operate under gubernatorial authority within state jurisdiction without Posse Comitatus restrictions. The NG MP units are the only NG units trained, skilled, and experienced in providing law enforcement. They offer civilian authorities a unique, specialized capability not available from other NG units.

SUPPORT TO HUMANITARIAN ASSISTANCE AND DISASTER RELIEF

11-12. Humanitarian assistance and disaster-relief operations provide emergency assistance to victims of natural and man-made disasters and other endemic conditions. Natural disasters include earthquakes, floods, and hurricanes. Man-made disasters include riots and civil conflicts. Humanitarian assistance and disaster relief conducted by US forces alleviate urgent nonmilitary needs of a populace until the appropriate civilian agencies can provide the needed services without military support. These types of operations include, but are not limited to—

- Conducting searches and rescues.
- Evacuating, sheltering, sustaining, and protecting dislocated civilians.
- Making food, medical care and treatment, and other civilian-aid programs available.
- Reducing opportunities for criminal activity and restoring L&O.
- Maintaining other operations needed to ensure the well being of the affected population.
- Providing relief operations for man-made disasters. These operations are characterized by civil conflict, and they entail all of the requirements for other disasters but focus on the restoration of L&O.
- Providing force protection and security.

11-13. MP support can be an important asset in all disaster-relief operations. The MP are trained and equipped for decentralized operations. MP teams operate in highly mobile vehicles equipped with radios, which are invaluable in disaster-relief operations. Each MP platoon, company, and battalion headquarters is equipped with high-frequency radios. These capabilities allow the MP (even when widely dispersed) to maintain centralized communications; to provide L&O support; and to assist in search, rescue, and evacuation efforts. Because of their decentralized operations and communications equipment, the MP can also play a vital role in notification, ADC, and access and egress control. They can deploy with disaster-assessment teams to assist in advising and developing a disaster-assistance plan. The MP can also deploy to an affected area to—

- Assist in restoring order.
- Search for, rescue, and evacuate victims.
- Provide personnel and vehicular circulation controls.
- Provide ADC.
- Conduct detention operations.
- Provide physical and area security to protect life and property immediately following a disaster.
- Protect fire and emergency-services personnel.
- Gather, record, and report information.
- Provide force protection and security.

11-14. MP (I/R) battalions can support civil-affairs units by establishing I/R facilities to shelter, sustain, protect, and account for disaster-relief victims. These units can also train and assist government agencies (US and foreign) or police in the operation of facilities for dislocated civilians and detainees.

11-15. The MP possess capabilities critical to relief in man-made disasters characterized by civil conflict. The MP have the essential training and skills needed for relief operations that focus on restoring civil order. Restoration of civil order demands—

- **Objectivity and neutrality.** The objective of civil-disturbance control is to restore L&O, not to impose punishment or suppress peaceful dissent. Government forces quelling civil disorders must be emotionally objective and politically neutral. MP law-enforcement training stresses objectivity and neutrality, an important attribute for forces employed in stability or support operations.
- **Minimal use of force.** The application of more than minimum force may be detrimental to mission success. The MP are particularly suited for these situations, having been trained on the technique for the use and measured application of minimum forces to accomplish mission goals.
- **Public acceptance.** Government forces must avoid actions that will alienate the populace. Any action offensive to community sensitivities must be avoided. The MP receive training in crowd-control operations, interpersonal communications skills, and minimum use-of-force techniques—skills that are critical in gaining public acceptance.
- **Positive image.** In many disturbance operations, demonstrators and government forces compete for the approval and sympathy of the general population. The MP's assist-and-protect image provides the government with a distinct advantage in such situations.
- **Threat awareness.** Government forces must pay attention to and try to learn the organization, intentions, and activities of demonstrators. Equal attention must be paid to how the community will respond to the demonstrator's actions and the government's response. The MP interact daily with the populace, placing them in a position to develop criminal information and to determine the collective attitude of the populace.
- **Effective coordination.** Disturbance operations often involve many agencies from various levels of government. Thorough, timely, and coordinated preparation, planning, training, and execution are essential for mission success. MP civil-disturbance training (including planning and coordination measures) and their communications assets enhance coordination, particularly during mission execution.
- **Operational unity.** Multiple control forces representing different agencies with varying jurisdictional authority can pose C^2 problems. When unity of command cannot be achieved, operational unity must be achieved by colocating operational centers, integrating communications, and delineating organizational responsibilities. The MP are well trained in police and military operations, possessing a sound understanding of both civilian police and military measures, legal constraints, and effective coordination.

11-16. The MP operational tasks supporting man-made disasters vary according to the nature of the disaster and the operational environment. For example, in CONUS where the commitment of military forces to civil disturbances is considered a last resort and military involvement is limited by law, nonfederalized NG MP units have long been employed by civil authorities for their unique capabilities. MP forces can—

- Isolate and contain an area with barricades, roadblocks, and perimeter patrols. The MP can also assist civil-affairs units by enforcing pass and identification systems and public-utility controls.
- Provide mobile patrols to enhance the security of high-priority targets (such as buildings, utilities, and services that are critical to the community's economic and physical well being).
- Monitor, disperse, contain, or limit crowd movement through—
  - Observation points and patrols.
  - The communication of interest and intent to the crowd.
  - The channelization or diversion of the crowd.
  - Gaining the cooperation of the dissident leadership.
  - The proclamation of the illegal nature of the crowd's actions.
  - Show of force.
  - The use of crowd-control formations.
  - The apprehension of crowd members and leaders.
  - The application of minimum-force measures.
- Establish area control to prevent looting, to protect businesses and other likely targets, and to prevent arson. The MP accomplish this mission with saturation patrolling, including vehicle and foot patrols. With augmentation, the MP can perform air and water patrols. They also enforce populace control measures (such as ordinances to prevent gathering, permits to gather, restrictions on circulation, restrictions on the interference with government and public functions and personnel, restrictions on possessing weapons, and other measures instituted by civil authorities).
- Neutralize special threats (such as snipers and bombings) that are highly dangerous to both government forces and the community. Snipers pose a particularly grave danger in disaster-relief operations. The MP can take immediate protective actions, secure an area, isolate the threat and, when authorized, use SRTs to assist civilian or HN authorities in apprehending or neutralizing the sniper. The MP are trained to perform the necessary actions when an explosive device is discovered. In such an event, they can secure and evacuate an area, organize search teams, and isolate the site.

11-17. The MP can also support in rescue and recovery operations, which may include rescuing US or foreign nationals. Such operations are also employed to locate, identify, and recover sensitive materials deemed critical to national security. The operations may be conducted in benign or hostile environments. The MP provide OPSEC, and they process and account for civilians affected by rescue and recovery operations. The MWD teams can also help search for lost or hidden personnel. In stability and support operations, the MP can perform their police functions for rescue and recovery operations while keeping the perceived military signature low.

SUPPORT TO COUNTERDRUG OPERATIONS

11-18. Counterdrug operations are measures taken to disrupt, interdict, and destroy illicit drug activities. In a counterdrug role, the DOD may offer certain forms of support to HN counterdrug personnel, to US civilian law-enforcement agencies operating in a HN, to Department of State (DOS) counterdrug personnel, and to the Bureau of International Narcotics Matters. Without direction from the National Command Authority acting under constitutional or statutory authority, US forces engaged in counterdrug activities may not engage in direct law-enforcement activities (such as apprehension, search, seizure, and other similar activities). As discussed earlier, the Posse Comitatus Act authorizes specific DOD assistance in counterdrug activities, but US military forces usually may not participate directly in arrests, searches, seizures, and similar activities. Conversely, as discussed earlier, USC Title 32 authorizes state governors to use NG MP units with full authority to engage directly in law-enforcement activities. When authorized, US advisors may accompany counterdrug forces on operations.

11-19. MP operational contributions can include the use of force, drug identification, crime-scene protection, customs operations, surveillance of criminal activity, serious-incident reporting, and other related activities. Employing the MP for these operations ensures force suitability and avoids a mismatch in terms of unit capability and mission effectiveness.

11-20. Border-screening operations consist of three separate but related measures—mobile patrolling, the use of listening/observation posts, and airfield surveillance. These measures are part of the MP's AS and MMS functions used in tactical rear-area operations. The MP's operational effectiveness may be further increased through augmentation by a civilian law-enforcement officer who would be responsible for search, seizure, and arrest actions. Active-component units are permitted to assist in marijuana eradication in support of domestic law-enforcement agencies (DLEAs). Marijuana-eradication operations are an application of the MP AS, mobility, and L&O functions, thus providing realistic training while supporting an operation.

11-21. As discussed in previous chapters, the MP are well suited to augment customs operations. They routinely perform customs inspections for US forces in Korea and Germany. The NG MP regularly conduct vehicle, cargo, and container searches under the supervision of US Customs agents and at US ports of entry.

11-22. The USACIDC agents can provide support to counterdrug operations. The CID units identify the sources of illegal-drug distribution systems. When directed, the MP and the CID complement interagency counterdrug efforts to stop the production, flow, and distribution of drugs and to provide planning, training, equipment, and facilities to support domestic counterdrug operations. The CID also supports CRIMINTEL programs, internal security, and site surveys at borders where Army units will deploy. In addition to the aforementioned support, the DOD General Counsel has ruled that MWD teams can be loaned to DLEAs to assist in detecting smuggled contraband.

SUPPORT IN COMBATING TERRORISM

11-23. The Department of Justice (DOJ) is the lead federal agency for dealing with acts of terrorism within the US, its territories, and its possessions. Within the DOJ, the Federal Bureau of Investigation (FBI) has the lead. The FBI can train the police of friendly nations in antiterrorism and counterterrorism operations. Usually, US military forces act in a technical-advisory but not a law-enforcement capacity in combating acts of terrorism. It is often difficult to distinguish the acts of politically motivated terrorists from violent acts of criminals or individuals in society at large. In fact, all terrorist acts are criminal; however, they also may be part of an insurgency.

11-24. The MP are key players in the defensive planning, implementation, and control of combating-terrorism operations, whether in garrison or deployed to war or MOOTW. In whatever countries US forces operate, MP units conduct antiterrorism operations daily as part of their L&O and AS functions. The MP also respond to terrorist incidents. They help secure the area and search for evidence, and they help the lead agency. The MP provide advice and assistance in developing antiterrorism and counterterrorism programs for deployed US forces and DOS agencies and for police, paramilitary, and military forces of assisted nations (when so authorized). MP training, liaison, and joint patrols help strengthen HN-police organizations, improving their effectiveness and efficiency in combating terrorism. The MP also assist HN police in performing PRC operations.

11-25. The training of police (domestic and foreign) requires DA approval, and restrictions exist on some forms of training. Legal advice and coordination are a must before training can occur. The MP and the CID provide personnel and equipment for planning, advising, equipping, and training agencies to prevent, deter, and respond to terrorism. The training focuses on deterring terrorist acts such as bombings, kidnappings, assassinations, taking hostages, and hijackings. Related assistance may include training and equipment for explosive detection, management of hostage situations, physical security, protection of key personnel, and defensive/evasive driving methods. For counterterrorism operations on US military installations, the MP provide highly trained, specially equipped SRTs for situations requiring the selective use of force. Additionally, CID agents are highly trained in hostage negotiation, which increases the commander's options before resorting to force.

11-26. The MP and the CID assist in counterterrorism information-gathering activities through liaison and joint patrols with HN police. These activities are part of the PIO function. The MP have extensive contacts with civilian police, who in turn have contacts with the populace. These contacts can help surface criminal information concerning terrorist plans, recruitment, and support structures sooner than military sources. The CID can provide commanders with CRIMINTEL on a daily basis. At the same time, MP antiterrorism support includes measures to reduce the vulnerability of people and property. Such measures include—

- Personal awareness.
- Personal-protection operations and techniques.
- Crime prevention.
- Physical-security programs.

11-27. The MP and CID crime-prevention programs employ proactive measures to protect people and their property and US and HN property. Crime prevention is a continuous process of planning, implementing measures, evaluating effectiveness, modifying measures, and developing further information. The primary crime-prevention tool is the crime-prevention survey. It is used to examine the physical features of various critical sites. The MP physical-security programs help identify, reduce, eliminate, or neutralize conditions favorable to criminal, terrorist, and insurgent activities. The key measures used to accomplish this are physical-security inspections and surveys.

SUPPORT TO NONCOMBATANT EVACUATION OPERATIONS

11-28. Noncombatant evacuation operations (NEOs) relocate civilians who are endangered in a foreign nation. The NEOs are classified as benign (unopposed) or hostile (opposed). MP roles in support to NEOs differ accordingly. In a benign environment, NEOs are usually conducted with the help and full cooperation of the affected nation, with little or no opposition to evacuation. The use of force is limited to self-defense (as outlined by the ROE) and protection of the evacuees. In a hostile environment, NEOs are opposed by forces with specific intentions of preventing or destroying the operation. In this scenario, the MP will follow the ROE as established by higher headquarters.

11-29. MP operations in NEOs closely parallel EPW/CI evacuation operations. The objective of both is to safely collect, secure, protect, transport, process, and account for people, whether they are EPWs or noncombatants. However, due to the dissimilarities of these groups, the techniques employed differ.

11-30. The MP can secure selected facilities, reception and transfer points, or other locations required to collect, process, and evacuate people. They can provide route reconnaissance and mobile security teams to escort HRP, DOS personnel, evacuees, and sensitive material. If needed, MP units assist civil-affairs personnel with the movement and control of dislocated civilians. The MP may support civil affairs by setting up and operating evacuation sites, holding areas for foreign nationals denied evacuation, and reception and processing stations. This support can be provided by MP CS units for a short duration and by MP I/R units for a prolonged duration. Additionally, the MP will establish and maintain liaison with HN police to ensure a mutual understanding of jurisdiction, to coordinate efforts, and to provide criminal and tactical intelligence. In hostile environments requiring the employment of combat forces, the MP would be employed in their traditional CS role.

SUPPORT TO SECURITY-ASSISTANCE OPERATIONS

11-31. Security-assistance operations are one of the main tools of US foreign policy. A critical concern in these operations is the HN's ability to plan and manage its own defense resources. In security-assistance operations, the US provides defense materiel, military training, and defense-related services through grants, loans, credit, or cash sales. When authorized, the MP support security assistance with military and police training through the International Military Education and Training Program (IMETP). This program provides instruction and training to foreign military and civilian personnel on a grant/aid basis. Many foreign civil law-enforcement agencies request training in the following areas:

- Basic police procedures (such as patrolling and crime-scene protection).
- Physical security.
- Corrections.
- Civil-disturbance operations.
- Customs operations.
- Traffic control.
- Use of force.

11-32. The MP possess the expertise and experience to advise, train, and assist a HN's military and police forces. The MP's participation in the IMETP includes—

- Formal and informal instruction of foreign students.
- Technical education and tactical training with applicable training aids.
- Assistance to foreign police and military elements by MP mobile training teams (MTTs).
- Providing MP units or small, modular, functionally specific teams to support security-assistance needs. These teams can train HN police in both field and institutional facilities.
- Support to nation-assistance operations.

11-33. Nation-assistance operations support a nation's effort to promote development, ideally by using its national resources. This is achieved through education and the transfer of essential skills to the HN. The goals of such operations are to—

- Promote long-term regional stability.
- Develop sound and responsive democratic institutions.
- Develop supportive infrastructures.
- Promote an environment that allows for orderly political change and economic progress.

11-34. The MP support nation-assistance operations with training and joint MP and HN-police operations. The MP know and understand police operations, military operations, and C^2 procedures. Consequently, the MP are ideal for training the law-enforcement personnel in both field and institutional settings. The MP's nation-assistance goals are to—

- Enhance L&O in democratic societies.
- Improve efficiency and effectiveness.
- Promote the proper usage and maintenance of equipment.
- Establish a sound training base for police.

- Standardize procedures that enhance combined police operations.
- Promote friendship and goodwill toward the US.
- Preclude the need for US military interventions to counter acts of violence.

11-35. Combined police operations reinforce training and provide HN police with mobility, security, and communications to operate more effectively while conserving critical personnel resources. The MP provide support through training and the following:

- Law-enforcement, security, and criminal-information support to the HN police force (to include L&O operations and administration in a democratic society).
- Patrol and desk operations.
- Circulation-control operations.
- PRC and civil-disturbance operations.
- MWDs.
- Physical-security operations.
- Personnel-security operations.
- Area- and route-security operations.
- Counterdrug operations.
- Antiterrorism operations.
- Mass immigration operations.
- Customs and border operations.
- Confinement operations.
- Crime prevention.
- Community and police relations.
- C^2 of police operations.

11-36. When conducting nation-assistance operations, as well as other operations, the MP will often be tasked to support civil-affairs operations. Close coordination with the G5/S5 and liaison with the appropriate HN representatives have added emphasis to these types of operations. The scope of civil-affairs operations is influenced by the economic, social, psychological, and political background of the country and its people. Civil-affairs personnel will coordinate the MP's effort to restore stability, contribute to national development, and promote support for the host government. Civil-affairs personnel assist in preventing civilian interference with military operations and coordinate all other military and civil affairs (such as community relations, PRC, civil defense, and community- and area-security programs).

11-37. The MP can conduct ADC measures to assist the HN in the aftermath of a natural or man-made disaster. They help rescue and evacuate the injured for medical care, and they issue food, water, and essential supplies (See FM 41-10).

SUPPORT TO PEACEKEEPING OPERATIONS

11-38. Peacekeeping operations (PKOs) are military operations conducted with the consent of the belligerents in a conflict. The PKOs are intended to maintain negotiated truces and to facilitate diplomatic solutions. The US may participate in a PKO under the auspices of an international organization, in cooperation with other countries, or unilaterally. A PKO usually occurs in an ambiguous situation in which there is extreme tension and the possibility of violence. This violence may range from terrorism and sabotage to minor military conflicts involving known and unknown belligerents. MP forces may serve as an element of a UN peacekeeping force or as part of a multinational observer group. The overall operational control of a peacekeeping force is exercised by a multinationally staffed military peacekeeping command. However, the commander of each assigned national force retains command integrity of his unit.

11-39. The basic MP force structure and appropriate augmentation are situation dependent. The size and composition of the MP forces are determined by diplomatic negotiations. Personnel spaces are allocated based on the US contingents' respective missions. The principle of consent affects the composition and function of the peacekeeping force. Consent applies to the degree of interest the disputing countries have when contributing their forces to participate in the peacekeeping effort.

11-40. Each PKO is unique but is generally characterized by constraints in the use of force. In a PKO, force is normally limited to self-defense, the defense of others, and the protection of national security. Therefore, the soldiers in a PKO must possess the skills required for warfare, but they must also be trained so that they are equally able to act as intermediaries. In this aspect, the MP are a distinct asset. They are experienced in a case-by-case approach to the use of force. They are experienced in using discretion and in the de-escalation of potential violence. The MP training in EPW operations are especially useful in handling EPWs if the peacekeeping force supervises the exchange of prisoners of war.

11-41. The peacekeeping force employs a combination of the following techniques to control the potential for violence inherent to PKOs:

- Observation is the peacekeeping force's primary responsibility and basic requirement. It provides timely and accurate reports of any suspicious situation, incident, or occurrence.
- Information gathering is a continuous requirement. Peacekeepers must be constantly alert to what takes place around them and to any changes or inconsistencies in the behavior, attitude, and activities of military and civilian personnel.
- Surveillance and supervision operations help oversee the implementation of agreements. The operations require restraint, tact, and patience.
- Complaint investigations must be thorough and objective, and they must result in a fair assessment.
- Negotiation and mediation skills are often required of peacekeepers. If peacekeepers can resolve minor problems at the lowest level, they can prevent major issues from rising, which may prevent increased tensions or the resumption of fighting.
- Patrolling (mounted and dismounted) among the population publicizes presence. It also serves as a means of gathering information, informally enhancing supervision, and investigating complaints.

11-42. The above techniques are compatible with MP battlefield functions. Operationally, the MP conduct route and area R&S, operate mounted or dismounted patrols, operate listening/observation posts, provide humanitarian assistance, investigate possible terrorist and criminal acts and cease-fire and sanction violations, and gather information. The MP provide a flexible, wide-ranging means of information dissemination. Additionally, they provide crucial support for force protection. The MP and the CID employ active and passive measures to deter and defeat threats directed at service members, their families, DOD civilians, facilities, equipment, and very important persons (VIPs). The MP use emerging technologies in force protection to conduct security, detection, and identification operations and to warn of intruders.

11-43. As part of a PKO, the MP conduct crowd- and riot-control operations, including the extraction of mob leaders. The MP are an ideal force for controlling antagonistic masses engaged in rioting, looting, and demonstrating. The MP understand how to make the transition from the lower end of the use-of-force spectrum to the use of lethal force, if so required or directed. Finally, the MP deploying with peacekeeping forces can provide early-on force protection, including headquarters and initial aerial port/ seaport of debarkation (APOD/SPOD) security. The MP's mobility, firepower, and communications provide critical reconnaissance, information-collection, and response-force capabilities and acceptability not found in combat forces.

11-44. When conducting crowd- and riot-control operations in a PKO or peace-enforcement operation (PEO), nonlethal weapons (NLWs) are an additional means of employing force for the purpose of limiting the probability of death or serious injury to noncombatants or belligerents. However, the use of lethal force must always remain an inherent right of individuals in instances when they, their fellow soldiers, or personnel in their charge are threatened with death or serious bodily harm. The NLWs add flexibility to the control of disturbances by providing an environment where military forces can permissively engage threatening targets with limited risk of noncombatant casualties and collateral damage (see FM 90-40).

11-45. The use of lethal force (employed under the standing ROE) will never be denied. At no time will forces be deployed without the ability to defend themselves against a lethal threat nor will they forego normal training, arming, and equipping for combat. Nonlethal options are a complement to, not a replacement for lethal force. They seek to expand a proactive response across the range of military operations.

11-46. The decision to use NLWs against an adversary during a confrontation should be delegated to the lowest possible level, preferably to the platoon or squad. However, this requires that all personnel (not just leaders) have a clear understanding of the ROE and the commander's intent.

11-47. Commanders and public-affairs officers must be prepared to address media questions and concerns regarding the use and role of NLWs. They must be prepared to address the role of NLWs, and they must make it clear that the presence of NLWs in no way indicates abandoning the option to employ deadly force in appropriate circumstances.

11-48. The PEOs are in the high end of the PKO's spectrum. These operations are just short of combat and require a different operational approach. MP operations to support PEOs correspond to MP battlefield functions supporting combat forces. The PEOs often pose special operational considerations for dealing with dislocated civilians. The MP provide PRC operations and I/R, when required.

ORGANIZATIONS AND CAPABILITIES

11-49. MP units are well suited for stability and support operations given their unique capabilities and low force signature. The MP brigade, battalion, company, and I/R units may be tailored for stability and support operations based on their unique METT-TC.

MP BRIGADE TF

11-50. In essence, the organization and capabilities of the MP organization in stability and support operations is the same as those for other operations. However, its augmentation is tailored based on METT-TC unique to stability and support operations and the commander's needs. The MP brigade has been the centerpiece of successful MP deployment during previous stability and support operations. The MP brigade provides comprehensive planning, C^2, and A/L support to assigned and attached subordinate units. The brigade also has the command and staff experience to coordinate and supervise HN security and law-enforcement activities.

11-51. The MP brigade headquarters provides senior MP leadership and the communications capabilities required to direct a variety of functional elements—CSS, PSYOP, public affairs, and combat-arms augmentation—in support of joint or unilateral TF operations. At authorized level of organization (ALO) 1, the MP brigade headquarters has an SJA section to coordinate the legal and jurisdictional issues inherent to stability and support operations. The brigade headquarters also has an S5 section and a liaison capability to coordinate support and assistance to HN agencies, friendly forces, other US forces, DOS personnel, and nongovernmental or private volunteer organizations.

11-52. The MP brigade TF (MPBTF), with its robust capabilities and low force signature, is well suited for the unique mission requirements and sensitivities of stability and support operations. The MPBTF can be employed as part of a combat TF in high-threat, opposed-entry operations or as the lead TF for a low-threat, unopposed-entry operation. The following are examples of previous MPBTFs and the operations in which they participated:

- **Operation Provide Comfort.** The 18th MP Brigade deployed to northern Iraq to provide humanitarian assistance and protection to the Kurds.
- **JTF Los Angeles.** The 49th MP Brigade (NG) provided civil-disturbance assistance to Los Angeles in the aftermath of the Rodney King trial.
- **Operation Hawkeye.** The 16th MP Brigade and units of the 89th MP Brigade provided security, protection, and humanitarian

assistance to the population of Saint Croix in the aftermath of Hurricane Hugo.

- **JTF Andrew.** The 16th MP Brigade provided humanitarian assistance in the aftermath of Hurricane Andrew.
- **Operation Sea Signal.** The 89th MP Brigade deployed to Guantanamo Bay, Cuba, to provide humanitarian assistance to Haitian migrants.
- **Operation Restore Democracy.** The 16th MP Brigade deployed to Haiti to provide humanitarian assistance.

11-53. Early deployments of the MPBTF headquarters or a mission-tailored portion of the headquarters positions a senior, effective MP C^2 headquarters to respond to all MP requirements. It can assist in the coordination with the media, governmental and nongovernmental agencies, and other forces, and it can coordinate civil-military aspects of the operation. The MP brigade headquarters can assimilate and employ additional forces without degrading operations. The augmentation of brigade headquarters is driven by METT-TC. With the brigade headquarters in place, subordinate or supporting units can be significantly expanded without the loss of C^2. In essence, little or no loss in operational momentum occurs when the C^2 element deploys first.

11-54. The MP LRP section of the MP brigade headquarters colocates and works with the higher headquarters' G3/Operations Directorate (J3) plans element. This allows the MP LRP to incorporate MP operations effectively into higher-level planning, and it enables an efficient transition to field operations. Also, because in peacetime this section functions as a PM cell, it provides the PM with efficient and effective technical support during stability and support operations. As the operations approach termination or enter sustainability and fewer forces are required, the brigade headquarters can redeploy, leaving a battalion TF in charge of MP forces.

MP BATTALION TF

11-55. The MP battalion (CS) provides command, control, and support for assigned MP companies (CS). MP battalions (CS) are best employed as part of a TF—either as part of a combat TF (high threat, opposed) or a support TF (low threat, unopposed). With sufficient augmentation, the battalion can provide the logistical, administrative, personnel, and operational support needed to lead a diverse TF. During stability and support operations, an MP battalion TF may be placed under the OPCON of a brigade-size TF (MP or non-MP) or a division TF. If placed under the OPCON of a division TF, both the division PM and the MP battalion commander will synchronize their efforts to maximize the employment of available divisional and nondivisional MP assets in support of the division commander's concept of operations.

11-56. As with the MPBTF, the MP battalion's contributions to stability and support operations are not a new concept. The following are examples of previous battalion TFs and the operations in which they were involved:

- **Cuban Refugee Movement.** The 519th, 716th, 720th, and 759th MP Battalions deployed for humanitarian-assistance operations.
- **Operations Restore Hope and Continued Hope.** The 720th MP Battalion provided humanitarian assistance in Somalia.

- **Operation Sea Signal.** The 720th MP Battalion deployed to Guantanamo Bay, Cuba, to conduct humanitarian-assistance operations for Haitian migrants. Later, the 716th and 759th MP Battalions also deployed to Guantanamo Bay to perform like functions for the Cubans.

MP COMPANY IN STABILITY AND SUPPORT OPERATIONS

11-57. The MP company is the MP Corps's primary mission performer. It executes a wide variety of missions. In these operations, MP companies—

- Provide mobility support for both vehicles and personnel.
- Provide AS, including security for critical personnel, sites, cargoes, and railways.
- Secure, safeguard, shelter, and control detainees and dislocated civilians.
- Restore order in civil disturbances and conduct PRC operations.
- Carry out limited L&O operations.
- Provide liaison, coordination, and training for all aspects of law enforcement to other agencies.
- Conduct police-intelligence collections and disseminate information to the populace, friendly forces, and other HN agencies.
- Support force protection and security.

11-58. The MP company possesses robust mobility, lethality, and communications. Each of the MP companies' three-man teams are equipped with either a high-mobility, multipurpose, wheeled vehicle (HMMWV), an up-armored HMMWV, or an armored security vehicle (ASV); an AN/VRC 91 radio; a Light-Vehicle Obscuration Smoke System (LVOSS); an MK-19 grenade machine gun; a squad automatic weapon (SAW); two M-16 or M-4 rifles; an M203 grenade launcher; and a 9-millimeter pistol as a side arm for each team member. In addition to conducting combat operations, the MP company commander has the capability to conduct limited L&O operations (desk and investigative operations). With a tremendous operational reach, the MP company will displace its platoons as far as possible to accomplish all of the MP functions.

MP I/R UNITS IN STABILITY AND SUPPORT OPERATIONS

11-59. The MP I/R units are specifically equipped and trained to support, safeguard, account for, and provide proper and humane treatment for EPWs/CIs, US military prisoners, or dislocated civilians. When conducting dislocated-civilian operations, the MP I/R units support and interact with civil-affairs and PSYOP units who are responsible for monitoring and handling dislocated civilians. Should more than one category of personnel have to be supported, the MP I/R units will establish separate compounds for each to preclude contact with other populations. When operation of the facility is going to be more than a short-term endeavor, the intent is to involve (as appropriate) NGOs, IHOs, HN authorities, and dislocated civilians in the administration of the facility. The tasks performed by MP I/R units for stability and support operations are basically the same as those for war, but the operations and procedures are modified according to the operational environment and the categories of people to be housed in the facility (see FM 19-40).

11-60. Criminal detainees (who are detained against their will) are subject to controls and procedures that closely parallel those used for EPWs/CIs. The primary emphasis is on guarding and preventing escape until their disposition is determined by recognized civilian judicial/law-enforcement authorities. When the operation is conducted outside of the US and its territories, criminal detainees are treated according to established legal procedures.

11-61. Dislocated civilians (who are being assisted rather than detained) are provided aid, shelter, and protection. The emphasis is on protecting them from harm by natural forces or hostile personnel. A special category of personnel arises when the MP I/R unit is required to house personnel that are dislocated civilians but who must be detained against their will. Such is the case of mass migrants who flee their countries and find themselves under US custody while formal proceeding guidance is being developed. In this case, MP I/R units must be sensitive to the situation and attempt to strike a balance between security, shelter, protection, and detention procedures.

11-62. Military detainees (who are detained against their will) are subject to controls and procedures that closely parallel those used for EPWs/CIs. In stability and support operations, US forces detain opposing-force military personnel. Without a formal declaration of war, opposing military personnel who are captured cannot be categorized as EPWs. However, they are afforded many of the EPW's rights and privileges under the Geneva convention. Such a situation occurred during Operation Just Cause when PDF personnel were detained by US forces.

11-63. In an operational environment in which hostile groups are engaged against one another, an I/R facility may be set up to protect one group from another. In this case, the facility's purpose is to shelter, sustain, account for, and protect its dislocated civilians from the hostile group. The MP I/R units and other supporting units concentrate on providing AS to protect the facility from direct fire. Other MP or combat forces provide protections beyond the direct-fire zone. Accountability for the dislocated civilians is coordinated with the SJA and civil affairs. They focus on maintaining a record of the people in the facility and their physical conditions. In operations in which no hostile groups are engaged (such as natural disasters), the I/R facility may be set up to provide shelter, sustain, and account for personnel. There may not be a need for external security personnel.

11-64. The MP I/R C^2 structure for stability and support operations is based on METT-TC. The nature and complexity of the mission, the number and types of detainees, and the operational duration should be considered. At one end of the operations spectrum, an MP brigade (I/R) may be required; while at the other end, an MP battalion (I/R) may be assigned to an MP brigade (CS).

Chapter 12

Force Protection

There is no more important responsibility than force protection. It is an inherent command responsibility and must be fully integrated into every unit's mission. A commander must continually review his unit's force-protection posture and crosswalk it with current and changing policy and threat levels. Force protection demands the personal involvement of commanders to ensure the best possible security consistent with the threat to personnel and mission-essential resources.

AR 525-13

Force protection consists of those actions that prevent or mitigate hostile actions against DOD personnel (to include family members), resources, facilities, and critical information. It coordinates and synchronizes offensive and defensive measures to enable the joint force to perform while degrading opportunities for the enemy. It includes air, space, and missile defense; NBC defense; antiterrorism; defensive-information operations; and security to operational forces and means.

OVERVIEW

12-1. The MP contribute significantly to the Army's overall force-protection effort. They are directly responsible for executing certain security-related measures (law enforcement and physical security) and for providing support to others (HRP security and antiterrorism), all of which contribute to force protection. Commanders use the Army's risk-management methodology and established security programs and capabilities within US forces to help prioritize and implement force-protection measures.

SUPPORT TO FORCE PROTECTION

12-2. MP support to force protection is accomplished under the leadership of the respective command's PM. It is accomplished mainly through an active role in physical security, HRP security, law enforcement, and antiterrorism.

PHYSICAL SECURITY

12-3. Physical security is an integral part of security of forces and means. Physical-security measures deter, detect, and defend against threats from terrorists, criminals, and unconventional forces. These measures include—

- Fencing and perimeter standoff space.
- Lighting and sensors.

- Vehicle barriers.
- Blast protection.
- Intrusion-detection systems and electronic surveillance.
- Procedural measures such as security checks, training and awareness, property accountability/inventory requirements, physical-security inspections of mission-essential or vulnerable areas (MEVAs), and physical-security surveys of installations.

12-4. These physical-security measures are not exclusively peacetime measures. They are designed to protect US personnel and equipment regardless of the mission or the geographical location.

12-5. The MP support the commander's security of forces and means programs by—

- Controlling or monitoring installation, base, or base-cluster access or entrance points.
- Monitoring intrusion-detection systems and providing a response force.
- Conducting physical-security inspections.
- Conducting perimeter security or site surveillance.
- Recommending the placement of walls, berms, gates, or barriers around designated MEVAs, high-value areas, or perimeters.
- Supporting the commander's risk-analysis effort.
- Conducting roving patrols, checkpoints, or roadblocks.
- Performing other physical-security measures as required by the commander.

HRP SECURITY

12-6. The HRP security program provides additional security to designated individuals (and their families) whom by virtue of their rank, assignment, symbolic value, vulnerabilities, or location or a specific threat are at a greater risk than the general population. This security program is applicable across the full spectrum of military operations.

12-7. The MP support the commander's HRP security measures by—

- Conducting protective-service details for HRP (normally conducted by USACIDC special agents).
- Responding to duress alarms and intrusion-detection systems.
- Conducting a PSVA (performed by the USACIDC).
- Employing MWDs.
- Coordinating with local, state, and federal agencies as well as joint, multinational, and HN law-enforcement agencies.
- Performing other HRP security measures as required by the commander.

LAW ENFORCEMENT

12-8. The Army's law-enforcement program supports the security of forces and means through the prevention, detection, response, and investigation of crimes. The MP, the USACIDC, and DOD police play a crucial role in the law-enforcement program. They provide a visible deterrent against a broad portion of the threat spectrum, investigate crimes, and provide an initial response force to threat incidents. As with the previous programs, the law-enforcement program is applicable anywhere across the spectrum of military operations.

12-9. The MP support law-enforcement operations by—

- Providing liaison teams with local, state, and federal agencies; HN police; and joint and multinational agencies.
- Employing SRTs and hostage-negotiation teams (normally provided by the USACIDC).
- Providing traffic enforcement, MSR regulation enforcement, and other route-control measures.
- Employing MWDs.
- Conducting MP investigations (criminal and noncriminal).
- Conducting patrolling, AS, and surveillance measures.
- Implementing applicable threat-condition (THREATCON) measures.
- Conducting and implementing other law-enforcement measures as required by the commander.

ANTITERRORISM

12-10. Terrorism is characterized as the unlawful use of violence or the threat of violence to coerce or intimidate a government or a society. Protection against a terrorist threat requires both an offensive counterterrorism capability and a defensive antiterrorism program. The primary focus of antiterrorism is on training and awareness and thorough planning. Antiterrorism is not a stand-alone program. It leverages other force-protection-related programs (such as physical security and law enforcement) to provide much of the physical defense against terrorism.

12-11. The MP support the commander's antiterrorist program by—

- Collecting CRIMINTEL (within the limits of AR 190-45).
- Responding to terrorist threats and incidents.
- Performing aggressive patrols and R&S operations.
- Detecting suspicious activities.
- Enforcing joint and cooperative agreements with the FBI, the USACIDC, MI, and other pertinent agencies (home or abroad).
- Conducting information collection and dissemination.
- Protecting critical assets, facilities, and personnel.
- Performing other antiterrorism measures as required by the commander.

Appendix A

Metric Conversion Chart

This appendix complies with current Army directives which state that the metric system will be incorporated into all new publications. Table A-1 is a conversion chart.

Table A-1. Metric Conversion Chart

Metric to English			English to Metric		
Multiply	**By**	**To Obtain**	**Multiply**	**By**	**To Obtain**
Length					
Centimeters	0.0394	Inches	Inches	2.54	Centimeters
Meters	3.28	Feet	Feet	0.0305	Meters
Meters	1.094	Yards	Yards	0.9144	Meters
Kilometers	0.621	Miles (stat)	Miles (stat)	1.5609	Kilometers
Kilometers	0.540	Miles (naut)	Miles (naut)	1.853	Kilometers
Millimeters	0.039	Inches	Inches	25.40	Millimeters
Area					
Square centimeters	0.1550	Square inches	Square inches	6.45	Square centimeters
Square meters	10.76	Square feet	Square feet	0.0929	Square meters
Square meters	1.196	Square yards	Square yards	0.836	Square meters
Volume					
Cubic centimeters	0.610	Cubic inches	Cubic inches	16.39	Cubic centimeters
Cubic meters	35.3	Cubic feet	Cubic feet	0.0283	Cubic meters
Cubic meters	1.308	Cubic yards	Cubic yards	0.765	Cubic meters
Milliliters	0.0338	US liq ounces	US liq ounces	29.6	Milliliters
Liters	1.057	US liq quarts	US liq quarts	0.946	Liters
Liters	0.264	US liq gallons	US liq gallons	3.79	Liters
Weight					
Grams	0.0353	Ounces	Ounces	28.4	Grams
Kilograms	2.20	Pounds	Pounds	0.454	Kilograms
Metric tons	1.102	Short tons	Short tons	0.907	Metric tons
Metric tons	0.984	Long tons	Long tons	1.016	Metric tons

Appendix B

Doctrine and Training Literature

Users of this manual must understand the elements of doctrinal literature and their relationship to each other. The commonly used terms doctrine and tactics, techniques, and procedures (TTP) are interrelated and mutually supportive; yet, each term has its own usage, level of detail, and place in the hierarchy of doctrinal publications. Because there is a close interrelationship, publications overlap in what they provide to the user. Figure B-1 illustrates the relationship of these doctrinal terms and publications.

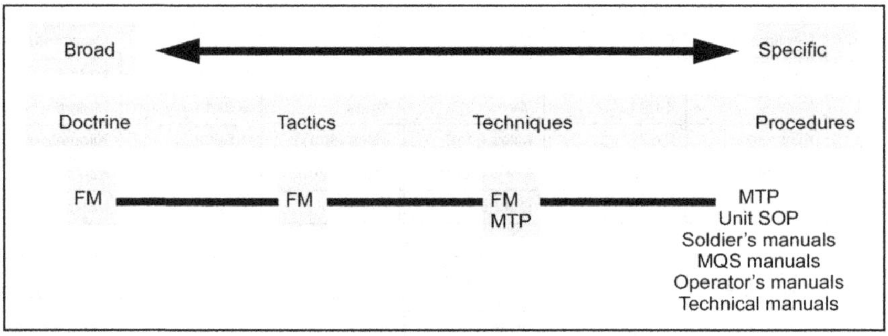

Figure B-1. Doctrinal and Training Literature

DOCTRINE

B-1. Doctrine is the body of fundamental principles by which the military guide their actions in support of national objectives. It is authoritative, but requires judgment in application. Doctrine provides the Army's fighting philosophy, establishes a common language, and structures the manner within which units plan and conduct combat operations.

B-2. Doctrine is not absolute; it is applied by the leader to meet the circumstances of the situation being faced. Doctrine provides for this flexibility with broad fundamentals and principles for conducting operations. Leaders well grounded in doctrine understand how to synchronize the elements of combat power and how to apply TTP using the assets available to accomplish the mission.

B-3. Doctrine is described in FMs. FM 3-0 defines the systems of war fighting practiced by the US Army. FMs 100-15 and 71-100 expand on this foundation. This manual describes the CS doctrine of the MP; it is the capstone manual for MP operations. It defines the role, operational requirements, functions, and C^2 relationships of MP units. This manual is not oriented on a specific organizational structure. Leaders at all levels apply this doctrine to their unique organizational and operational situations and use various tactics and techniques to accomplish the doctrine.

TACTICS

B-4. Tactics describe how the leader carries out doctrine (relating to the arrangement of forces for battle). Tactics include the—

- Employment of units in combat (to accomplish a mission assigned by the commander).
- Ordered arrangement and maneuver of units in relation to each other and to the enemy in order to use their full potential (a description of how the commander should arrange his forces and maneuver to accomplish a mission).

B-5. This is the part of doctrinal literature used for training in preparation for combat. Like doctrine, tactics presented in manuals are applied with judgment by the leader. The FMs prescribe "how to fight" and the mission training plans (MTPs) describe "what to train."

TECHNIQUES

B-6. Techniques are the methods of performing any act, especially the detailed methods used by troops or commanders in performing assigned tasks. Techniques describe the basic methods of using equipment and personnel. Techniques give details on how commanders actually carry out assignments. They improve the force's efficiency by ensuring the uniformity of action or by ensuring that the actions of various individuals and elements complement those of other individuals or elements.

B-7. The FMs and MTPs provide a description of techniques using the personnel and equipment available in actual organizations. These techniques show (at a more detailed level) how to use available assets to carry out tactics.

PROCEDURES

B-8. Procedures are the lowest level of detail. They address "how to" at the task level. Procedures are a particular course or mode of action that describe how to perform a certain task. Procedures include the—

- Standing methods used by units to accomplish tasks.
- Operating procedures for weapons and equipment.
- Crew drills.
- Staff actions and coordination requirements.
- Methods of target engagement by direct- and indirect-fire weapons.

B-9. Procedures are building blocks of individual- and collective-task accomplishment that serve as the foundation of tactics and techniques. Procedures are explained in unit standing operating procedures (SOPs), MTPs, soldier's manuals, operator's manuals, technical manuals (TMs), and similar publications.

Appendix C

Battlefield Workload Analysis

With minor adjustments in the capability block, the BWA can be used to determine MP requirements for the division. A BWA work sheet is shown in Figure C-1. A separate matrix is prepared for each EAC and corps MP brigade. Figure C-2, page C-3, is a sample of a BWA matrix for a 3-division corps MP CS company.

Mission/Function[1]	Capability[1]	Workload[2]	Rqmt	HN[18]	US Rqmt	Risk[19]	Mission Rqmt
Divisions supported[3]	1						
MMS	360 km MSR and ASR						
Refugees on control routes[4]	150,000 per day						
AS—corps rural terrain[5]	2,000 sq km						
AS—EAC rural terrain[5]	3,000 sq km						
AS—urban terrain[5]	800 sq km (urban)						
AS—USAF main opns base[5]	1,000 sq km/USAF MOB						
AS—convoy escort[6]	# tms per convoy	# convoys					
AS—ammo security[7]	1.3 modular ammo co						
	2 modular ammo co						
AS—pipeline security[8]	400 km						
AS—EAC pipeline co[8]	3						
AS—trains[9]	7						
AS—small critical sites[10]	12						
AS—major HQ security[11]	1						
AS—EAC port security[12]	0.66 of port						
AS—combat operations	When required, assets are diverted from other missions.						
L&O—incident response[13]	75,000 nondivision troops						
I/R—US prisoner detention[14]	700						
I/R—EPW escort/corps[15]	1,900 walking						
	2,500 by vehicle	Min: 1 plt					
	3,800 by train						
I/R—EPW holding/corps[16]	2,000	Min: 1 plt					
Other assigned missions[17]	Mission	Force size					
		Totals[20]					
Battalion HHD requirement = 0.199 x number of companies[21]							

Figure C-1. Sample BWA Work Sheet

NOTE: Workload ÷ capability = doctrinal requirement - HN offset = US requirement - acceptable risk = mission requirement

[1]Mission/capabilities are as stated in section I of TOEs 19477L000 and 19677L000. The unit capability for each mission is what the unit can accomplish when performing no other mission (for example, if it is only doing MSRs, the unit can cover 360 kilometers in a 24-hour period). Unless otherwise specified, the capabilities are the same at corps and EAC.

[2]Workload is the amount (MSR or area) or number of supported units/facilities.

[3]Corps support to each division is covered by allocating one MP company per supported division. Any additional support requirements should be based on a detailed division PM/MP mission analysis and included under other assigned missions.

[4]Unless specifically tasked, the Army normally relies on HN support to manage refugees. Redirecting refugees from MSRs and keeping them off of MSRs is part of (as is straggler control and TCP) the mission and MSR workload. When assigned the mission, the capability is for control of refugees on established refugee control routes that are in addition to the MSRs.

[5]Compute each square kilometer mission requirement separately. Do not compute the same terrain in more than one computation (for example, the area around the Air Force's main operating base (MOB) is not included with either rural or urban square kilometers).

[6]Convoy-escort capability is not stated in the TOE and is an AS function rather than an MMS function. Determine the average number of teams per mission and the average number of simultaneous missions. Multiply the number of teams required per mission by 0.028, then multiple the result by the number of simultaneous missions to determine the total requirement. State the number of teams per convoy and the number of convoys, then convert the force size to a 3-decimal-place company equivalent. Each additional team, squad, and platoon equivalent equals 0.028, 0.083, and 0.25 respectively.

[7]Security for modular ammo companies (ASPs) supporting divisions is three platoons and is normally conducted directly behind the division (corps forward). However, METT-TC may require the ammo units to move into the DRA. In this case, corps units should be attached to the division for the duration. Security for corps storage areas (CSAs) and theater storage areas (TSAs) is one MP company per two ammo companies. Security is provided on an area-support basis in the form of screening, with limited fixed posts.

[8]The EAC petroleum terminal/pipeline operating company manages the EAC pipeline that is normally laid adjacent to MSRs. Pipeline security is inherently provided by the MP operating on the MSRs. Unless the pipeline is not near the MSR, MP support is only allocated against the EAC petroleum terminal/pipeline operating company's bulk storage facilities. Security is provided on an area-support basis in the form of screening, with limited fixed posts.

[9]Train security is normally an EAC mission. This capability equates to simultaneously providing five MP teams to each of seven trains. Actual workloads and requirements are computed in the same manner as a convoy escort.

[10]Small critical facilities are normally defined as key MSR bridges and tunnels and communication nodes that may require additional security.

[11]The major headquarters are corps, COSCOM, TSC, unified command and, if directed, the JTF. The ASCC depends on the MP to provide security for two CPs. The ASCC requirement is an additional three platoons to support a second CP.

[12]The requirement for one seaport is 1.5 companies.

[13]Teams provide incident response, crime-scene security, and traffic control. They also assist L&O detachment teams and CID investigators.

[14]For deliberate planning in support of an OPLAN, personnel to be confined will normally be held by either the unit or the EAC I/R MP battalion performing the confinement mission. This does not preclude a requirement for the echelon PM to operate a field detention facility until prisoners are released back to their units or evacuated to the EAC confinement facility.

[15]The primary mode is vehicle (back-haul doctrine). Regardless of the mode, a minimum of one platoon (0.25 requirement) per supported division is allocated for EPW escort from the division central collection point (DCCP) to the corps holding area (CHA). Because the actual number of EPWs is unknown, coupled with no time/distance study for each supported division, one platoon per division provides a minimum capability for continuous evacuation within the stated doctrinal timeline (evacuate to the CHA within 12 hours of arrival at the DCCP).

[16]At a minimum, one platoon (0.25 requirement) is allocated to operate the CHA (provides the capability to hold 500 EPWs at any given time).

[17]Add any additional continuous missions assigned by the echelon commander not covered above. Examples include VIP escort/security for other than the CG and the deputy commanding general (DCG) of major headquarters, screening exposed corps flanks, and increasing teams per mission/function which decreases doctrinal capability (for example, permanent TCPs or checkpoints). Each additional team, squad, or platoon equivalent equals 0.028, 0.083, and 0.25 respectively.

Figure C-1. Sample BWA Work Sheet (continued)

[18]For each mission/function, any workload either fully or partially accomplished by the HN is subtracted from the doctrinal mission requirement. Enter the MP company equivalent offset and subtract from the doctrinal requirements to determine the US requirement.

[19]For each mission/function, if the operational/MP commander will accept mission risk, subtract the MP company equivalent to determine the mission requirement.

[20]Add columns to determine the total doctrinal company requirement, HN offsets, US requirement, acceptable risk, and mission requirement to support the operation/AO. Round the requirement to the nearest whole number (0.4 or fewer rounds down, 0.5 or greater round up).

[21]Determine battalion HHD requirements by multiplying the company requirements by 0.199 and rounding.

Figure C-1. Sample BWA Work Sheet (continued)

Mission/Function	Capability	Workload	Rqmt	HN	US Rqmt	Risk	Mission Rqmt
Divisions supported	1	3	3.000		3.000		3.000
MMS	360 km MSR and ASR	1,050	2.917		2.917		2.917
Refugees on control routes	150,000 per day	25,000	0.167	0.167	0.000		0.000
AS—corps rural terrain	2,000 sq km	6,500	3.250		3.250	0.500	2.750
AS—EAC rural terrain	3,000 sq km						
AS—urban terrain	800 sq km (urban)	1,500	1.875		1.875	0.500	1.625
AS—USAF main opns base	1,000 sq km/USAF MOB	1,000	1.000		1.000		1.000
AS—convoy escort	5 tms per convoy	4	0.560		0.560		0.560
AS—ammo security	1.3 modular ammo co	3	2.308		2.308		2.308
	2 modular ammo co	1	0.500		0.500		0.500
AS—pipeline security	400 km						
AS—EAC pipeline co	3						
AS—trains	7						
AS—small critical sites	12	9	0.750		0.750	0.250	0.500
AS—major HQ security	1	2	2.000		2.000		2.000
AS—EAC port security	0.66 of port						
AS—combat operations	When required, assets are diverted from other missions.						
L&O—incident response	75,000 nondivision troops	45,000	0.600		0.600	0.300	0.300
I/R—US prisoner detention	700						
I/R—EPW escort/corps	1,900 walking						
	2,500 by vehicle	Min: 1 plt/div	0.750		0.750		0.750
	3,800 by train						
I/R—EPW holding/corps	2,000	Min: 1 plt/	0.250		0.250		0.250
Other assigned missions	PSD for Corps CofS	1 sqd	0.083		0.083		0.083
	Flank screen	1 plt	0.250		0.250		0.250
		Totals	20	0.177	20	1.3	19
Battalion HHD requirement = 0.199 x number of companies			4		4		4

SAMPLE

NOTE: Workload ÷ capability = doctrinal requirement - HN offset = US requirement - acceptable risk = mission requirement

Figure C-2. Sample BWA Matrix

Appendix D

Automated Information Systems in Support of Battle Command

As the MP Corps continues to advance into the twenty-first century, it must continue to pursue and stay in tune with the Army's initiative to acquire advanced technologies and operational concepts. This will give MP soldiers on the battlefield an information advantage that is critical for successful performance of their functions. The following information opens a window into the near future and explains some of the automation initiatives that will enhance the Army's and the MP Corps's soldiers.

ARMY INFORMATION SYSTEMS

D-1. The Global Command and Control System (GCCS) is the midterm solution and the bridge to the concepts outlined in the concept of command, control, communications, computers, and intelligence for the warrior (C4IFTW). The GCCS's core consists of the basic functions required by the war fighter to plan, execute, and manage military operations. The Army's selected C^2 system is the Army Battle Command System (ABCS).

D-2. The ABCS integrates the C^2 systems found at each echelon—from the ground-force component commander at the theater or JTF level to the individual soldier or the weapons platform. Whether ARFOR are deployed for land combat or are conducting peace operations, providing humanitarian assistance, or giving aid to civil authorities, the ABCS supports the mission by integrating the battle-space automation systems and communications that functionally link strategic and tactical headquarters. The ABCS is interoperable with joint and multinational C^2 systems at upper echelons across the full range of C^2 functionality, and it is vertically and horizontally integrated at the tactical and operational levels. The ABCS has three major components—

- The Global Command and Control System—Army (GCCS-A).
- The Army Tactical Command and Control System (ATCCS).
- The FBCB2.

GLOBAL COMMAND AND CONTROL SYSTEM—ARMY

D-3. The GCCS-A is the Army's component of the joint GCCS. It was built from application programs developed by the Army Worldwide Military Command and Control System [WMCCS] Information System (AWIS), namely the—

- Strategic Theater Command and Control System (STCCS).

- Theater Automated Command and Control Information Management System (TACCIMS).
- EAC portion of the Combat Service Support Control System (CSSCS) (CSSCS/EAC).

D-4. The GCCS-A is required in the MP LRP, S1, S3/NBC, and S4 sections within the MP brigade at EAC; in the Assistant Chief of Staff, G1 (Personnel) (G1), the G3, and the Assistant Chief of Staff, G4 (Logistics) (G4) sections within the MP command (I/R); and in the S1, S3, and S4 sections within the MP brigade (I/R).

ARMY TACTICAL COMMAND AND CONTROL SYSTEM

D-5. The ATCCS embodies the commander and his staff and the doctrine, procedures, and tools used to command and control forces on the tactical battlefield. The ATCCS's functional subsystems provide tools to support the C^2 process across the spectrum of low- to high-intensity conflicts worldwide; for MOOTW; and for combat, CS, and CSS functional areas at corps-and-below units. The ATCCS provides a common set of C^2 tools and functions for commanders that extends the ATCCS to the battalion level. The component's ATCCS's are the—

- Maneuver Control System (MCS).
- All-Source Analysis System (ASAS).
- Advanced Field-Artillery Tactical Data System (AFATDS).
- Air- and Missile-Defense Planning and Control System (AMDPCS).
- CSSCS.

D-6. The FBCB2 provides situational awareness and C^2 to the lowest tactical echelons. It facilitates a seamless flow of battle-command information and situation-awareness information across the battle space at brigade and below on Army platforms.

FORCE XXI BATTLE COMMAND—BRIGADE AND BELOW

D-7. The FBCB2 system, a battle-command information system, provides the technology to complete the ABCS information flow process from brigade to platform and across all platforms within the brigade TF. The FBCB2 enables commanders to remotely operate and maintain ABCS database connectivity (regardless of the command vehicle) and to digitally control and monitor their subordinate units' statuses and positions. Interoperable FBCB2 systems will enhance the battle commander's ability to have a detailed picture of the battlefield (situational awareness), to maximize force survivability, and to facilitate battlefield synchronization to reduce or destroy the enemy's capability to fight. The FBCB2 relies on integrating, processing, and moving information in a digital form. To support projections of future conflicts with an accompanying increase in the tempo of battle, the FBCB2 must be able to obtain information contained in databases from higher, lower, and adjacent elements without formally requesting that information through command channels. The war fighter requires the ability to pull data from all battlefield-system databases to keep current with the rapid changes caused by the dynamics of battle. The FBCB2 is required for divisional MP company commanders, PMs, platoon leaders, teams, and squads.

MANEUVER CONTROL SYSTEM

D-8. The MCS is the ABCS's horizontal and vertical integrator. The MCS will enable commanders and their staffs to collect, coordinate, and act swiftly on near-real-time battlefield information. It will quickly, accurately, and automatically move tactical information on the battlefield. This allows a commander to readily access information and display current situation reports (SITREPs) that assess enemy strength and movement as well as the status of friendly forces. The MCS will then aid the battle staff in rapidly disseminating the commander's orders. It is intended to provide two major functional capabilities—maneuver functional-area control and force-level information. The MCS is the ATCCS's functional subsystem that—

- Supports the combined-arms force commander directly by providing automated support for planning, coordinating, controlling, and using maneuver functional-area assets and tasks.
- Coordinates and synchronizes supporting arms in conducting operational planning, field operations, and training.

D-9. The MCS is required for division PM operations sections, company CPs at echelons corps and below (ECB), MP battalion commanders, S3 sections at corps, MP brigade commanders, and corps MP LRP and S3 sections.

ALL-SOURCE ANALYSIS SYSTEM—REMOTE WORK STATION

D-10. The ASAS—remote workstation (RWS) subsystem will operate within the existing intelligence and electronic warfare (IEW) battlefield operating system (BOS) and ABCS architectures to provide automation support to G2/S2 intelligence staff elements and other designated intelligence organizations from the maneuver battalion through EAC. It will operate at the security level appropriate to its echelon, mission, and functions. It assists the G2/S2 in providing IEW support to the commander and other staff elements in planning, coordinating, and conducting current and future tactical operations. It interfaces with the sensitive compartmented information (SCI) analysis and control element (ACE) subsystem to provide operational guidance and to exchange IEW information and products. It also operates as the technical control portion of the IEW C^2 node of the ABCS to provide current IEW and enemy-situation information to the force-level information database for access and use by ABCS users. Two RWSs are also provided to the brigade analysis control team (ACT) of the DS MI company to process data input from DS collection systems, to filter intelligence data from broadcast systems, and to pass tactically relevant enemy-situation information or situation-development data to the brigade S2's RWS. The ASAS-RWS is required in the S2 section of the CID group, the S2/S3 section of the CID battalion, the G2 section of the MP command (I/R), the S2 section of the MP battalion (I/R), and the S2 section of the MP battalion and brigade at all echelons.

COMBAT SERVICE SUPPORT CONTROL SYSTEM

D-11. The CSSCS will function as an integral part of the ATTCS. It is designated as an information system with the capability to rapidly collect, analyze, and disseminate logistical, medical, and personnel information to accelerate and improve decision making by CSS and force commanders. It also enables the rapid exchange of C^2 information with other battlefield functional areas (BFAs) through automated interfaces. The system will process unclassified through secret information. The CSSCS will be employed in all divisions, corps, EAC, separate brigades, and armored cavalry regiments. The MP require CSSCS functionality in all S1 and S4 sections in battalion or higher organizations except where the GCCS-A is required as stated above.

TRANSPORTATION COORDINATOR AUTOMATED MOVEMENT SERVER

D-12. The Transportation Coordinator Automated Movement Server (TC-AIMS II) is used by transportation agents and deploying units of each service to automate the processes of planning, organizing, coordinating, and controlling deployment and redeployment operations worldwide during peace as well as contingencies. The TC AIMS II will link all DOD-component unit movements with the installation transportation officer/traffic-management officer (ITO/TMO). This will provide the capability to plan and execute deployments rapidly to meet CINC-required arrival and departure dates. The TC AIMS II will also provide the ITO's/TMO's day-to-day traffic management functions and the theater transportation managers with the capabilities to use theater common lift assets effectively. The TC AIMS II will also allow CSS organizations and supporting commands to sustain the force and perform effective movement control, coordination, distribution, and sustainment once in theater. This is also defined as transportation movement and support of DOD personnel and cargo during all phases of military operations in all environments (including RSOI and battlefield operations). The TC AIMS II's functionality is required at all division PM operations and S3 sections within MP battalions and brigades to facilitate MSR regulation enforcement. Additionally, MP companies, battalions, brigades, and commands require the TC AIMS II to plan, organize, coordinate, and control deployments and redeployments.

GLOBAL COMBAT SUPPORT SYSTEM—ARMY

D-13. The GCSS-A is an integrated logistical support system with accompanying hardware that will connect the service's supply, finance, transportation, maintenance, and personnel databases. Under this system, the Army will integrate 13 Legacy tactical logistics systems and will eventually interface with the service's wholesale logistics systems as well as the consolidated logistics systems of the other services and the DOD. The system is critical for modernizing automation, for reducing logistical footprints in theaters, and for expanding war fighters' capabilities. The system will also replace the Standard Army Management Information System (STAMIS). The GCSS-A is required at all MP division company CPs and MP battalion and brigade S4 sections. Additionally, it is required at all supply functions within MP detachments that support I/R missions.

MILITARY POLICE AUTOMATED SYSTEMS

D-14. In an effort to maintain an information advantage and to acquire advance technologies and operational concepts, the MP Corps has acquired the following systems:

- The Military Police Management Information System (MPMIS) and its subsystems
- The CID's automated systems and their subsystems.

MILITARY POLICE MANAGEMENT INFORMATION SYSTEM

D-15. The MPMIS is a DA-mandated reporting system that does not interface with the ABCS. The MPMIS provides automated tools for MP records management and reporting. It reduces the administrative burden on the PM's staff. It increases the awareness of problem areas by managing crime information. The MPMIS consists of the following elements:

- **Offense Reporting Systems Windows (ORS-2W).** The ORS-2W provides automation support to track and report complaints, criminal incidents, and traffic incidents. It incorporates the CRC, the federally mandated National Incident-Based Reporting System (NIBRS), the DOD-mandated Defense Incident-Based Reporting System (DIBRS), the military community (MILCOM), and the installation's data requirements. The ORS-2W's functionality is required at PM activities at all echelons of command within the Army.
- **Correctional Reporting System (CRS-3).** The CRS-3 provides automation support to track and report inmates and corrections staff at all Army-controlled correctional facilities.
- **Security Management System (SMS).** The SMS provides automated support to assist physical-security inspectors and force-protection officers in reporting inspections and surveys. It tracks intrusion-detection devices and MWD training and health requirements.
- **Vehicle Registration System (VRS).** The VRS provides installations with an automated means to track vehicle, weapon, and bicycle registrations.
- **Prisoner-of-War Information System (PWIS-3).** The PWIS-3 provides the automated means of tracking an EPW's status from capture through release. This system is required at all EPW management activities.

CID AUTOMATED SYSTEMS

D-16. The Headquarters (HQ) USACIDC mandates CID automated systems. Similar to the MPMIS, it does not interface with the ABCS. The system facilitates data-collection efforts within the command. The CID's automated systems consist of the—

- **Automated Criminal Investigative Reporting System (ACIRS).** The ACIRS supports management and data-collection functions at the CRC and HQ USACIDC. It supports the following:
  - Maintaining investigative data.
  - Conducting agent investigative activities.
  - Reporting investigative data.
  - Managing the investigative processes.
  - Conducting criminal-activity analyses.

- **CRIMINTEL.** Data sources for CRIMINTEL include MP reports, CRIMINTEL reports and bulletins, reports of investigation, crime-prevention surveys, case files, collection folders, raw-data folders, significant data files, and target-analysis files. The CRIMINTEL interfaces with the ACIRS, but it does not share CRIMINTEL products.

Appendix E

Police Information Assessment Process

The PIAP differs somewhat from the tactical IPB. While the IPB allows the maneuver commander to see the effects that a variety of factors may have on his forces, the PIAP looks beyond this approach to determine how other relevant information may impact his forces. This information may be obtained from police or criminal actions or incidents encountered during the performance of MP functions.

IMPLEMENTATION

E-1. The PIAP is not a substitute for the IPB. MP leaders must continue to rely on the IPB and to use its estimates as a starting point for the PIAP. However, the PIAP is a dynamic and continuous cycle that complements the IPB (Figure E-1). The following paragraphs represent the doctrinal approach to implementing the PIAP. These steps provide the basic foundation to the process. They are not exclusive; the successful development of the PIAP will depend on the MP leader's ability to apply the process to his specific environment, METT-TC, and the commander's priorities.

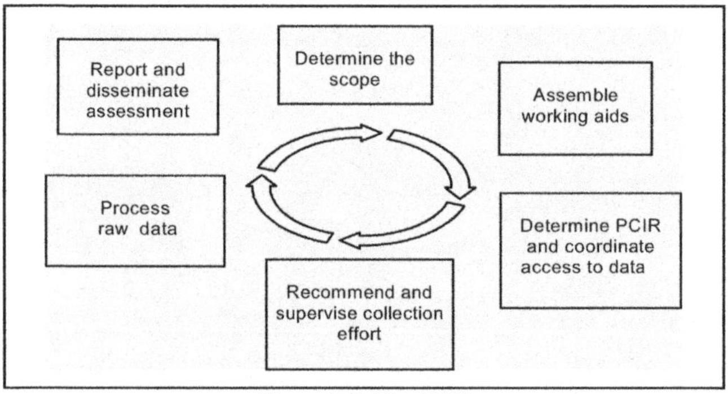

Figure E-1. The PIAP

DETERMINE THE SCOPE OF THE PIAP

E-2. While determining the scope of the PIAP might sound simple, it is one of the most crucial elements of the entire process. It entails identifying what you want to be the end result of your PIAP. To help determine the scope of the PIAP, MP leaders must perform the following subtasks:

- **Conduct a mission analysis.** A detailed mission analysis will ensure that all available resources are focused toward one goal. The more specific the mission is, the more focused and productive the effort. This involves more than just looking at the mission statement, the commander's intent, and the concept of operations. It involves translating the mission and situation into easily understood, manageable goals. The following are examples of these missions:
  - Identifying criminal threats from local national to US Army logistics operations within the port of Mogadishu, Somalia.
  - Determining the impact of gang violence on US forces operating in the vicinity of Camp Bondsteel, Kosovo.
  - Determining how rioters' violence will affect force protection before deploying forces to Los Angeles, California.
  - Determining the status of public service agencies (such as the police, the fire department, and the emergency medical services [EMS]) in Homestead, Florida, after the city was struck by Hurricane Andrew.

NOTE: As you can see from the examples, these missions are quite specific. In all likelihood, it would be necessary to establish several different PIAP missions to adequately protect Army elements operating in the MP's AOR.

- **Prioritize missions.** MP leaders must prioritize each PIAP mission since MP resources will probably be limited. The commander's intent is the key factor for determining priorities. When the commander's most important mission has been determined, MP leaders must decide how they can contribute to the overall success of that mission.
- **Determine the AO.** Generally, the AO is a geographical area (including the airspace above) usually defined by lateral, forward, and rear boundaries assigned to a commander. MP leaders must know where major units are located and how the characteristics of the location may impact ongoing operations. For example, units in direct contact with the enemy or isolated from the local population are less likely to be the target of a PIAP mission than those units adjacent to great population centers or in rest and relaxation centers.
- **Determine key terrain.** Key terrain is any locality or area that affords a marked advantage to the retaining side.
- **Determine the AOI.** An AOI is the area of concern to the commander. This includes the area of influence and the areas adjacent thereto and extending into enemy territory and to the objectives of current or planned operations. A thorough understanding of the AOI may help predict potential threat actions. For example, how far are local agitators and ringleaders willing to travel to interfere with US operations in Pristina, Kosovo? Would they go to other US compounds?
- **Determine required information products/sources.** The desired end-state product must be decided. Should it be a report? An updated SITREP? An upgrade in force-protection measures? Or a combination of products? It should also be determined who can provide the best

information (the CID; MI; the local police chief; or joint, interagency, or multinational agencies).

ASSEMBLE WORKING AIDS

E-3. During this step, MP leaders assemble the necessary graphic and nongraphic aids and information that will help them accomplish their mission. The CID and the MP may form an analysis team to track and analyze police information that might assist tactical operations or population-control operations. Working aids include, but are not limited to—

- **Maps.** Besides the standard military maps that depict the military AOs, the MP must attempt to procure indigenous maps of AOs. Detailed city and street maps will prove to be invaluable even if they are in a foreign language.
- **Language aids.** Phrase books and translation dictionaries cannot take the place of intensive language training, but they are better than nothing. These are essential even if the MP have assigned interpreters.
- **Open-source information.** This can cover an extensive range of invaluable information, but it should be available from a variety of sources. Examples include—
  - Demographic information on the HN population.
  - Societal information on the HN.
  - The locale and disposition of HN police forces.
  - Historical crime data of the area.
  - Environmental information (terrain and weather factors).
  - Local newspaper articles.
  - Internet sources on all of the above.

NOTE: Sources for this type of information can come from intelligence estimates provided by the higher headquarters' S2/G2, other military units, or the HN.

- **OPSEC information.** This information includes the disposition of US forces (to include MP, MI, CID, and other services' investigating units) deploying or being used within the AO.
- **Other working aids.** These may include items such as computers, printers, boards, and acetate that will be necessary depending on the missions the MP are undertaking.

DETERMINE INFORMATION REQUIREMENTS AND COORDINATE ACCESS TO DATA

E-4. Once the mission analysis is completed, the next step is to determine if there are information gaps. An information gap is a missing piece of information that is critical to the analysis. MP leaders must identify and prioritize the gaps using the time, the available resources, and the commander's intent. Once an information gap is identified, it becomes PCIR. Examples of PCIR are as follows:

- Where is the nearest police station, and how did they handle a certain problem?

- What is the composition of the two gangs operating in our AO, and how can they influence US operations?
- Why is river-crossing point Bravo a danger to US crossing forces? Why are local rioters disrupting only US operations on the east bank?

E-5. The next step is to determine if PCIR can be answered with data already available. To do this, the available military and nonmilitary sources must be identified. Some sources may include—

- **USACIDC.** The USACIDC is responsible for programs such as combating terrorism, CRIMINTEL, personal-security assessment, crime analysis, and LOGSEC threat assessment. Some of the information gathered by the USACIDC (although releasable under existing controls and restraints) may help with the PIAP mission. In fact, USACIDC special agents may already be working in the AO and may be available to help the MP with the collection effort.
- **MI element.** Like the USACIDC, the MI element operating in the AO may have come across a piece of information (police or criminal) that may help accomplish the PIAP mission. The MP must make daily contact with the S2/G2 and coordinate access to invaluable information. As with the USACIDC, available intelligence information may only be releasable under strict controls and restrictions imposed by the commander.
- **Other MP units.** Coordination with an adjacent MP unit may prove to be beneficial, especially if the other MP unit has conducted a similar PIAP mission or has conducted operations in the AOI.
- **HN law-enforcement agencies.** Local law-enforcement agencies will provide the essential populace knowledge otherwise not available to US forces. Chiefs of police, sheriffs, and other key community leaders may help MP leaders fill the information need of the PIAP mission.
- **Joint, interagency, and multinational forces.** Close and effective liaison with these elements will result in quick access to information that can produce the same benefits as the agencies listed above.

NOTE: If, after consulting with all available sources, the PCIR is not resolved, the MP leader must decide the best collection effort to solve the PCIR.

RECOMMEND AND SUPERVISE THE POLICE/CRIMINAL COLLECTION EFFORT

E-6. The collection effort is the means by which specific PCIR will be met. This implies developing a collection strategy, tasking specific collectors, and supervising the collection effort.

- **Develop a collection strategy.** After a through study of the availability, capability, and disposition of the potential collecting resources, MP leaders select which asset is better suitable to perform the mission. Are organic MP the best collectors for this mission? Or do we need to request support from USACIDC units? What unique or organic capabilities do each bring to the effort? Which MP function is the best? Do we conduct MMS, AS, or L&O operations? Part of the collection strategy includes coordinating with the S2/G2, the SJA, the

CID, and other agencies before launching the collection effort. This coordination will eliminate duplication of effort, interference with an ongoing effort, or stepping out of legal limits.

- **Task or request specific collectors.** The missions that will be tasked to respective collectors must be determined. Information collectors can be tasked with more than one mission at a time. However, it is imperative that their tasks be prioritized based on mission requirements and time available. The appropriate tasking or request chain must be used to request an MP team, a platoon, or USACIDC special agents.
- **Supervise the collection effort.** The collectors must be provided with reporting guidelines. How often should they report? Should they report "no information?" Specific reporting instructions should be provided, including how, when, and where they report. A collection tasking chart should also be used (Figure E-2). As more and more collectors are being used, it becomes important to track their missions, capabilities, and success. A brief note on the reporting instructions of each collector should be included.

PIAP Missions	Unit Tasked	Echelon Employed	POC	Collection Strategy Employed	Date Collection Started	Location	Date Collection Ended
PCIR 1							
PCIR 2							
PCIR 3							
PCIR 4							

Figure E-2. Collection Tasking Chart

PROCESS POLICE/CRIMINAL RAW DATA

E-7. This processing step converts raw data into police intelligence through analysis. The key to processing the data is to understand it. For example, as collectors report raw data it may often appear to be meaningless information. Rarely will one collector receive all of the necessary information to answer the PCIR at one time. Instead, a variety of information must be fitted together to form a coherent, even if incomplete, picture.

E-8. Every piece of data is important. This is the cardinal rule of the entire process. Until the data is analyzed for reliability and compared with other data, all reports must be treated equally. This is important to remember, for the trend will be to use data that either fits your expectations or gives the most complete picture.

E-9. A report of "no activity" can be as important as reports with information. If the collector is capable of collecting the information and is actively trying to collect, reports of "no activity" can be critical. First, if there is no indication of a change, they can be used to reprioritize PIAP missions. Second, they can be used to retask the collectors to other areas. It is important to check the mission's original intent periodically to ensure that it is still valid and merits further efforts.

E-10. To determine the reliability of the data, the MP leader must evaluate the source. Does the information come from the police chief, local youth, or other military forces? One of the best ways to determine the reliability of the information is to have more than one collector reporting on the same PCIR. If more than one collector reports the same information, it can normally be presumed to be accurate.

E-11. Information can be analyzed in a variety of ways—chronologically, geographically, by the impact on the force, by the type of activity or the modus operandi, or by the association of participants. The MP leader must evaluate the information after it is analyzed to determine whether the information collected answers the PCIR or whether it helps accomplish the PIAP mission. If the information does not answer these questions, he must decide whether it is necessary to task other collectors, retask the same collectors, or wait until the situation develops.

REPORT AND/OR DISSEMINATE POLICE AND CRIMINAL ASSESSMENT

E-12. Reporting and disseminating is the most important aspect of the PIAP. Collection and analysis is valueless unless the data is communicated to commanders who can best use the information.

E-13. Reporting requirements must be established based on the unit's SOP and other established processes. MP leaders must decide if the information collected must be sent using an established priority format or if it can wait to be sent using SITREPs. Does the information collected affect the force-protection measures or does it immediately impact the tactical scenario? Do we need to alert units operating in the vicinity of AA Fox to avoid sector 2? Additionally, MP leaders must ensure that any information released follows existing constraints and regulations.

E-14. It is also advantageous to provide feedback to the collectors. Collectors often receive no feedback on their performance. Informing them as to which part of their information was vital and beneficial can often increase their productivity in future operations.

SAMPLE SCENARIO

E-15. The following example helps illustrate the implementation of all six steps of the PIAP:

E-16. While deployed in support of Operation Control Chaos, US forces assigned to TF Blue Thunder are tasked with stability and support operations in the city of Corbina. This city, although currently implementing a peace treaty, was the site of extreme civil unrest. Even though the overall situation is not as chaotic as it was before US intervention, there are some sectors that are still in turmoil. Common crimes (such as robberies, assaults, looting, and vandalism) still occur daily and there is fear that this condition may spread out of control throughout the city. The TF commander knows that in order to bring peace to the entire city, he must attempt to stabilize the situation in some of these bad sectors. He decides that the best way to deal with this situation is to assign a sector (sector 51) to an MP battalion in an attempt to control the situation. If the MP battalion succeeds, he may use the same concept throughout the city. He is aware of the five MP functions and how they can help bring this sector under some control. Under this framework, the TF commander issues the following mission: *On order, the MP battalion will conduct stability and support operations in the vicinity of sector 51, the city of Corbina, in support of TF Blue Thunder.* Upon receipt of this mission, the battalion commander gathered his staff and initiated the planning process. Before the MP staff could initiate the PIAP, they made contact with the S2/G2 and received an updated IPB estimate. Now that the IPB estimates are on-hand, the staff initiates the PIAP.

DETERMINE THE SCOPE OF THE PIAP

E-17. The staff performs the following steps to determine the scope of the PIAP:

- **Conduct a mission analysis.** The staff looks at the given mission, the higher echelon commander's intent, and the concept of operations. After considering all of these factors, they produce a restated mission—*On order, the battalion will conduct MP operations in sector 51 to determine the type of threat and to bring this sector under control in support of TF Blue Thunder.*
- **Prioritize PIAP missions.** Since there is only one mission, this subtask is not applicable.
- **Determine the AO and the AOI.** The staff determines that the AO is the entire sector 51 and that the AOIs are the sectors adjacent to 51.
- **Determine key terrain.** The staff determines that the key terrains are the market place, the plaza, and the city park. They select these areas because most of the incidents occurred in these populated areas. Stabilization in these key areas may be advantageous to mission success.
- **Determine required information products/sources.** The end result is the stabilization of sector 51. Minimize the crime incidents and restore L&O. Information should come from the local police chief, local community leaders, civil affairs, PSYOP, and the CID.

ASSEMBLE WORKING AIDS

E-18. The staff assembles the following working aids in support of the PIAP:

- **Maps.** The staff acquires local maps that detail the market place, the plaza, and the city park. Although the maps are in a foreign language, the staff is able to identify these three prominent features and surrounding areas.
- **Language aids.** In addition to dictionaries, the staff is able to get two translators from civil affairs.

- **Open-source information.** Open-source information includes a variety of encyclopedias and country books. Newspaper reports and historical crime data is available via the internet or through MI sources. The emphasis is in obtaining information on the history and composition of the local police force.
- **Classified intelligence.** The staff receives a briefing from the S2/G2 describing the operations of SOF units in the area.
- **Other working aids.** Computers, charts, templates, and acetate are available.

DETERMINE INFORMATION REQUIREMENTS AND COORDINATE ACCESS TO DATA

E-19. After a careful analysis of all available information, the following information gaps were turned into PCIR:

- Why is the police force not actively engaged in L&O?
- What is the major threat to civilians in each area?
- Are local gangs involved?
- Is their local judicial system in place?
- Where are prisoners housed?

E-20. The staff consults with the USACIDC, MI, other MP units, and other sources but determines that the information provided is not clear or sufficient. Since the PCIR are not resolved, the staff decides to recommend that the commander initiate a collection effort.

RECOMMEND AND SUPERVISE THE POLICE/CRIMINAL COLLECTION EFFORT

E-21. The staff performs the following steps to recommend and supervise the police/criminal collection effort:

- **Develop a collection strategy.** Upon coordinating with the S2/G2, the SJA, and the CID, the MP commander decides that the best collectors will come from within organic assets. He decides that he will employ the MP to go out and seek the needed information. The battalion commander will employ a company commander to interview the local police chief and an MP/interpreter will patrol around the designated AOI. The MP commander will need additional interpreters to help him with his collection effort.
- **Task or request specific collectors.** The MP staff requests an interpreter through the normal chain of command. The TF G3 assigns six interpreters for a period of 90 days.
- **Supervise the collection effort.** All collectors are instructed to report not later than (NLT) 1800 every day. The report will be in the form of an "end of day" outbrief to the S2 and the battalion commander. A collection tasking chart (Figure E-3) will also be completed.

PROCESS POLICE/CRIMINAL RAW DATA

E-22. After receiving all of the above information, the S2 concludes that the inability of the police force to do their job has had a domino effect on all areas of peace and order. This information is confirmed through other reports from information collected by the CID (through their own efforts) and the CA. The information also reveals that the gangs have intentions of attacking US forces if they interfere with their "business." The S2 determines that although the situation may seem to be chaotic, once the locals are able to trust the police forces, everything should return to normal.

REPORT AND/OR DISSEMINATE POLICE AND CRIMINAL ASSESSMENT

E-23. Since the collection effort identifies a potential threat to US forces, the commander decides to notify higher headquarters and increase his own force-protection measures. He requires MP elements patrolling in designated areas to be no less than a squad. If patrolling dismounted, a HMMWV or an ASV will trail the squad. The MP commander uses the information provided during the PIAP to develop his implementation plan. This plan includes joint patrols, training, and construction of a joint MP/local national police station.

	PIAP Mission: Bring Sector 51 Under Control							
PCIR	**Unit Tasked**	**Echelon Employ-ed**	**POC**	**Collection Strategy Employed**	**Date Collection Started**	**Location**	**Date Collection Ended**	**PCIR Result**
Why is the police force not actively engaged in L&O operations?	A Co	2 teams	CPT Smith	Direct interview	31 Jan 00	Sector 51, Police HQ	2 Feb 00	The police force does not know how to interact with the population.
What is the major threat to civilians in each area?	A Co	6 teams	1LT Thomas	Patrol and interview	31 Jan 00	Market place, plaza, and park	7 Feb 00	Muggings, robberies, and stabbings are the major threats to civilians.
Are local gangs involved?	A Co	2 teams	CPT Smith	Direct interview	31 Jan 00	Sector 51, Police HQ	2 Feb 00	Two rival gangs exist.
Is a local judicial system in place?	A Co	2 teams	1LT Jones	Direct interview	31 Jan 00	Courthouse	3 Feb 00	A judicial system exists, but the judge is afraid to sentence anyone.
Where are the prisoners housed?	A Co	4 teams	1LT Lucas	AS and MMS	31 Jan 00	Local prison	2 Feb 00	The local prison is abandoned.

Figure E-3. Sample Collection Tasking Chart

Glossary

1LT first lieutenant

A/L administrative and logistics

AA assembly area

AATF air-assault task force

ABCS Army Battle Command Systems

ACE analysis and control element

ACIRS Automated Criminal Investigative Reporting System

ACT analysis control team

ADA air-defense artillery

ADC area damage control

AFATDS Advanced Field-Artillery Tactical Data System

ALO authorized level of organization

AMDPCS Air- and Missile-Defense Planning and Control System

AO area of operation

AOE Army of Excellence

AOI area of interest

AOR area of responsibility

APOD aerial port of debarkation

APOE aerial port of embarkation

AR armored

AR Army regulation

ARFOR Army forces

AS area security

ASAS All-Source Analysis System

ASCC Army service component command

ASG area support group

ASP ammunition supply point

ASR alternate supply route

ASV armored security vehicle

ATCCS Army Tactical Command and Control System

attn attention

AWIS Army WMCCS Information System

BCT brigade combat team

bde brigade

BFA battlefield functional area

BLD brigade liaison detachment

bn battalion

BOS battlefield operating system

BRT brigade reconnaissance troop

BWA battlefield workload analysis

C^2	command and control
C^3	command, control, and communications
C4I	command, control, communications, computers, and intelligence
C4IFTW	command, control, communications, computers, and intelligence for the warrior
CAS	close air support
CASE	corps-area support element
CCIU	computer-crime investigative unit
cdr	commander
CG	commanding general
CHA	corps holding area
CI	civilian internee
CID	Criminal Investigation Division
CIDR	Criminal Investigation Division Regulation
CINC	commander in chief
CIP	Criminal Intelligence Program
cmd	command
CMOC	civil-military operations center
co	company
COA	course of action
CofS	Chief of Staff
commo	communications

COMMZ communications zone

conf confinement

CONUS continental US

COSCOM corps support command

CP command post

CPT captain

CRC Crime Records Center

CRIMINTEL criminal intelligence

CRS-3 Correctional Reporting System

CS combat support

CSA corps storage area

CSE corps support element

CSG corps support group

CSM command sergeant major

CSS combat service support

CSSCS Combat Service Support Control System

cust customs

CZ combat zone

DA Department of the Army

DC District of Columbia

DCCP division central collection point

DCG deputy commanding general

DIBRS Defense Incident-Based Reporting System

DISCOM division support command

div division

DLEA domestic law-enforcement agency

DMMC division materiel management center

DNA deoxyribonucleic acid

DOD Department of Defense

DOJ Department of Justice

DOS Department of State

DPM deputy provost marshal

DRA division rear area

DS direct support

DSA division support area

DSB division support battalion

DSE division support element

DSS decisive, shaping, and sustaining

DZ drop zone

EAC echelons above corps

EAD echelons above division

ECB echelons corps and below

EG escort guard

EMS emergency medical services

EPW enemy prisoner of war

ERP engineer regulation point

ES escort

FA field artillery

FBCB2 Force XXI Battle Command—Brigade and Below

FBI Federal Bureau of Investigation

Feb February

FEBA forward edge of the battle area

FIU field investigative unit

FM field manual

FOB forward operating base

FOIA Freedom of Information Act

FORSCOM US Army Forces Command

FSB forward-support battalion

FSC forward-support company

G1 Assistant Chief of Staff, G1 (Personnel)

G2 Assistant Chief of Staff, G2 (Intelligence)

G3 Assistant Chief of Staff, G3 (Operations and Plans)

G4 Assistant Chief of Staff, G4 (Logistics)

G5 Assistant Chief of Staff, G5 (Civil Affairs)

G6 Assistant Chief of Staff, G6 (Communications)

GAC ground-assault convoy

GCCS Global Command and Control System

GCCS-A Global Command and Control System—Army

GD guard

GO government organization

GS general support

HHB headquarters and headquarters battery

HHC headquarters and headquarters company

HHD headquarters and headquarters detachment

HMMWV high-mobility, multipurpose, wheeled vehicle

HN host nation

HQ headquarters

HRP high-risk personnel

HVA high-value asset

I/R internment and resettlement

IBCT initial/interim brigade combat team

IC information center

IEW intelligence and electronic warfare

IG inspector general

IHO international humanitarian organization

IMETP International Military Education and Training Program

IO international organization

IPB intelligence preparation of the battlefield

IRIC internment/resettlement information center

ISB intermediate staging base

ITO installation transportation officer

J3 Operations Directorate

Jan January

JFC joint-force command

JOA joint-operations area

JP joint publication

JRA joint rear area

JSOA joint special operations area

JTF joint task force

km kilometer(s)

L&O law and order

LID light infantry division

liq liquid

LOC lines of communication

LOGSEC logistics security

LRP	long-range plan
LSTA	logistics-security threat assessment
LVOSS	Light-Vehicle Obscuration Smoke System
LZ	landing zone
M/O	marksman/observer
M/S	mobility/survivability
MACOM	major Army command
MAJ	major
MCRP	Marine Corps reference publication
MCS	Maneuver Control System
MDMP	military decision-making process
mech	mechanized
MEDEVAC	medical evacuation
METL	mission-essential task list
METT-TC	mission, enemy, terrain, troops, time available, and civilian considerations
MEVA	mission-essential or vulnerable area
MFPU	major fraud procurement unit
MI	military intelligence
MILCOM	military community
min	minimum
MKT	mobile kitchen trailer

MLRS Multiple-Launch Rocket System

MMS maneuver and mobility support

MOB main operating base

MOOTW military operations other than war

MP military police

MPBTF military police brigade task force

MPMIS Military Police Management Information System

MQS military qualification standards

MSR main supply route

MTOE modified table of organization and equipment

MTP mission training plan

MTT mobile training team

MWD military working dog

NATO North Atlantic Treaty Organization

naut nautical

NBC nuclear, biological, and chemical

NCO noncommissioned officer

NEO noncombatant evacuation operation

NG national guard

NGO nongovernmental organization

NIBRS National Incident-Based Reporting System

NLT not later than

NLT nonlethal weapon

NWP Naval warfare publication

OCONUS outside the continental US

OPCON operational control

OPLAN operation plan

opns operations

OPSEC operations security

ORS-2W Offense Reporting Systems Windows

PASE port-area support element

PCIR police/criminal information requirements

PDF Panamanian Defense Force

PEO peace-enforcement operation

PI police intelligence

PIAP police information assessment process

PIO police intelligence operations

PIR priority intelligence requirements

PKO peacekeeping operation

plt platoon

PM provost marshal

POC point of contact

POD port of debarkation

POE port of embarkation

POL petroleum, oil, and lubricants

PRC populace and resource control

PSD protective security detail

PSU protective-service unit

PSVA personal-security vulnerability assessment

PSYOP psychological operations

pub publication

PWIS-3 Prisoner-of-War Information System

PZ pickup zone

R&S reconnaissance and surveillance

RAOC rear-area operations center

RC reserve component

ROC rear operations center

ROE rules of engagement

rqmt requirement

RSOI reception, staging, onward movement, and integration

RSTA reconnaissance, surveillance, and target acquisition

RWS remote workstation

S1 Adjutant (US Army)

S2 Intelligence Officer (US Army)

S3 Operations and Training Officer (US Army)

S4 Supply Officer (US Army)

S5 Civil-Affairs Officer (US Army)

S6 Communications Officer (US Army)

SALUTE size, activity, location, unit, time, and equipment

SAW squad automatic weapon

SCI sensitive compartmented information

sec section

SEP separate

SFC sergeant first class

SITREP situation report

SJA staff judge advocate

SMS Security Management System

SOC security-operations cell

SOF special-operations forces

SOFA status of forces agreement

SOI signal operating instructions

SOP standing operating procedures

SPOD seaport of debarkation

SPOE seaport of embarkation

SPOTREP	spot report
spt	support
sq	square
sqd	squad
SRT	special-reaction team
STAMIS	Standard Army Management Information System
stat	statute
STCCS	Strategic Theater Command and Control System
TA	theater Army
TAA	tactical assembly area
TAACOM	Theater Army Area Command
TAC	tactical
TACCIMS	Theater Automated Command and Control Information Management System
TAI	traffic accident investigation
TASE	TSC-area support element
TC-AIMS	Transportation Coordinator Automated Movement Server
TCF	tactical combat force
TCP	traffic-control post
TF	task force
THREATCON	threat condition
TM	technical manual

TMO traffic-management officer

tms teams

TO theater of operations

TOE tables of organization and equipment

TRADOC US Army Training and Doctrine Command

TRP traffic-regulation plan

TSA theater storage area

TSC theater support command

TSE tactical support element

TTP tactics, techniques, and procedures

UMT unit ministry team

UN United Nations

US United States

USACIDC US Army Criminal Investigation Command

USACIL US Army Criminal Investigations Laboratory

USAF US Air Force

USAMPS US Army Military Police School

USC US Code

USCG United States Coast Guard

USCS US Customs Service

USDA US Department of Agriculture

VIP very important person

VRS Vehicle Registration System

WMCCS Worldwide Military Command and Control System

Bibliography

AR 15-6. *Procedures for Investigating Officers and Boards of Officers.* 11 May 1988.

AR 190-12. *Military Police Working Dogs.* 30 September 1993.

AR 190-45. *Law Enforcement Reporting.* 30 September 1988.

AR 190-47. *The Army Corrections System.* 15 August 1996.

AR 195-2. *Criminal Investigation Activities.* 30 October 1985.

AR 525-13. *Antiterrorism / Force Protection (AT / FP): Security of Personnel, Information, and Critical Resources.* 10 September 1998.

CIDR 195-1. *Special Agent Operations Manual.* October 1994.

DA Form 2028. *Recommended Changes to Publications and Blank Forms.* 1 February 1974.

FM 3-0. *Operations.* To be published within the next 6 months.

FM 19-1. *Military Police Support for the AirLand Battle.* 23 May 1988.

FM 19-4. *Military Police Battlefield Circulation Control, Area Security, and Enemy Prisoner of War Operations.* 7 May 1993.

FM 19-10. *The Military Police Law and Order Operations.* 30 September 1987.

FM 19-40. *Enemy Prisoners of War, Civilian Internees and Detained Persons.* 27 February 1976.

FM 34-2. *Collection Management and Synchronization Planning.* 8 March 1994.

FM 34-130. *Intelligence Preparation of the Battlefield.* 8 July 1994.

FM 41-10. *Civil Affairs Operations.* 14 February 2000.

FM 63-4. *Combat Service Support Operations—Theater Army Area Command.* 24 September 1984.

FM 71-3. *The Armored and Mechanized Infantry Brigade.* 8 January 1996.

FM 71-100. *Division Operations.* 28 August 1996.

FM 90-13-1. *Combined Arms Breaching Operations.* 28 February 1991.

FM 90-40. *Multiservice Procedures for the Tactical Employment of Nonlethal Weapons (NLW) {MCRP 3-15.8; NWP 3-07-31; USCG Pub 3-07.31}.* 6 October 1998.

FM 100-1. *The Army.* 14 June 1994.

FM 100-5. *Operations.* 14 June 1993.

FM 100-8. *The Army in Multinational Operations.* 24 November 1997.

FM 100-10. *Combat Service Support.* 3 October 1995.

FM 100-15. *Corps Operations.* 29 October 1996.

FM 100-17-3. *Reception, Staging, Onward Movement, and Integration.* 17 March 1999.

FM 101-5. *Staff Organization and Operations.* 31 May 1997.

FM 101-5-1. *Operational Terms and Graphics (MCRP 5-2A).* 30 September 1997.

JP 3-10. *Joint Doctrine for Rear Area Operations.* 28 May 1996.

USC Title 18. *Crimes and Criminal Procedure.* 23 January 2000.

USC Title 32. *National Guard.* 5 January 1999.

Index

Foreword

We remain a nation at war in an era of persistent conflict, but we do not stand alone. Our nation has many multinational partners, equally committed to freedom, rule of law and stability. It is clear that we are stronger when we act with partners in today's operating environment. Therefore, security force assistance is no longer an "additional duty." It is now a core competency of our Army.

As distinctions separating war and peace blur and challenges to security increase, we must seek to help our multinational partners successfully confront their security challenges. Security force assistance builds our multinational partners' capability to defeat regular, irregular, and hybrid threats prevalent in an era of persistent conflict.

The two pillars of security force assistance are the modular brigade and Soldiers acting as advisors. Their tactical efforts build partner capability and improve the security situation. Over time, U.S. forces and partners foreign security forces collectively set the conditions to defeat common threats and ultimately achieve strategic success.

This manual addresses common characteristics and considerations for conducting security force assistance and clarifies what units and individual advisors must understand to work "by, with, and through" their counterparts. Since every situation and foreign security force is unique, units and individuals conducting security force assistance must carefully analyze the operational environment, especially the relationships of foreign security forces and their population. Conducting successful security force assistance requires adaptive units led by well-informed, culturally astute leaders. FM 3-07.1 provides commanders, staffs, and advisors with the common doctrine for achieving this goal.

Conducting foreign security assistance requires great skill in building relationships and "leading from behind." We must all increase our understanding of this important mission.

MARTIN E. DEMPSEY
General, U.S. Army
Commanding General
U.S. Army Training and Doctrine
Command

Field Manual
No. 3-07.1

Headquarters
Department of the Army
Washington, DC, 1 May 2009

Security Force Assistance

Contents

Figures

Preface

This field manual (FM) is the Army's doctrinal publication for security force assistance (SFA). It provides doctrinal guidance and direction for how U.S. forces contribute to SFA. It focuses on the brigade combat team (BCT) conducting SFA and advising foreign security forces. It is based on lessons learned from previous advising efforts and recent combat operations with a view to the future. It supports the Army Education System instruction on the theory and conduct of SFA.

The two primary audiences for this manual are leaders in BCTs conducting SFA and Soldiers assigned as advisors. The BCT is the largest fixed tactical organization and the key formation of the Army's modular organization. Army modularity allows commanders to add selective units to assist the BCT as it conducts SFA. The BCT and higher echelons provide the framework for advisors to function and accomplish the mission—developing foreign security forces and, when appropriate, working by, with, and through foreign security forces to achieve the desired end state of the host nation's internal defense and development. Thus, Soldiers assigned as advisors are the key element of SFA.

This FM provides the conceptual framework for conventional forces to conduct SFA within the construct of full spectrum operations, across the spectrum of conflict. It addresses SFA at operational and tactical levels.

Army doctrine is consistent and compatible with joint doctrine. FM 3-07.1 expands on fundamental principles for SFA expressed in FM 3-07. FM 3-07.1 uses text and concepts developed with North Atlantic Treaty Organization partners and ABCA (America, Britain, Canada, Australia, and New Zealand program armies).

Terms that have joint or Army definitions are italicized and the number of the proponent manual follows the definition. FM 3-07.1 will be the proponent manual for foreign security forces as a term.

FM 3-07.1 applies to the Active Army, Army National Guard/U.S. Army National Guard, and U.S. Army Reserve unless otherwise stated.

Headquarters, US Army Training and Doctrine Command is the proponent for this publication. The preparing agency is the Combined Arms Doctrine Directorate, US Army Combined Arms Center. Send written comments and recommendations on Department of the Army (DA) Form 2028 (Recommended Changes to Publications and Blank Forms) directly to: Commander, US Army Combined Arms Center and Fort Leavenworth, ATTN: ATZL-CD (FM 3-07.1), 201 Reynolds Avenue (Building 285), Fort Leavenworth, KS 66027-1352. Send comments and recommendations by e-mail to leav-cadd-web-cadd@conus.army.mil. Follow the DA Form 2028 format or submit an electronic DA Form 2028.

Introduction

Our strategy emphasizes building the capacities of a broad spectrum of partners as the basis for long-term security. We must also seek to strengthen the resiliency of the international system to deal with conflict when it occurs.

National Defense Strategy, June 2008

In an era of persistent conflict, the United States supports the internal defense and development of international partners, regardless of whether those partners are highly developed and stable or less developed and emerging. While many of these partners are nations, they can also include alliances, coalitions, and regional organizations. U.S. support to these partners ranges from providing humanitarian assistance to major combat operations. U.S. support includes conducting conflict transformation, bolstering partner legitimacy, and building partner capacity. A vital part of these three aspects of U.S. support is assisting partner security forces.

Security force assistance (SFA) is not new for Army forces. In fact, General George Washington's Inspector General of the Army acted as an advisor for Army forces. Baron Friedrick Wilhelm von Steuben instilled discipline and professionalism into an army that previously lacked formalized training. His 1779 *Regulations for the Order and Discipline of the Troops of the United States*, adapted from the Prussian army, formed the doctrinal backbone of the Continental Army throughout the Revolutionary War. Additionally, the lineage of the Army's operations field manual, FM 3-0, can be traced to this document. As a benefactor of advisors such as von Steuben, the Army has since undertaken what is called SFA several times throughout its history.

Army doctrine defines *security force assistance* as the unified action to generate, employ, and sustain local, host-nation, or regional security forces in support of a legitimate authority (FM 3-07). SFA is part of the FM 3-0 construct of full spectrum operations. Similarly, it is conducted across the spectrum of conflict or in any of the operational themes. SFA is normally part of a larger security sector reform effort, while in other instances, SFA is not tied to reform but to building partner capacity.

Three general situations exist in which SFA may occur: an internally focused bilateral relationship, an externally focused bilateral relationship, and a multilateral relationship. Consequently, SFA supports the appropriate partner's plans. When SFA supports a host nation, it also supports that host nation's strategy. If SFA supports a host nation's externally focused efforts, it must support the host nation's national security strategy. SFA may support regional security forces, such as those of the African Union or the Organization of American States. In these cases, SFA supports that organization's plans.

Organization

This FM primarily focuses on the brigade combat team and advisors. It also provides a wider doctrinal framework for SFA. It is organized in the following manner:

- **Chapter 1 focuses above the brigade combat team**. It provides the overall doctrinal context for SFA.
- **Chapters 2 through 6 focus on the brigade combat team level**. These chapters provide a framework for SFA, its considerations for the brigade combat team operations process, considerations for augmenting the modular brigade for security force assistance, its considerations for the unit employment, and its sustainment considerations.
- **Chapters 7 through 10 focus on the individual advisor**. These chapters discuss the advisor, advisor cultural and communication considerations, the advisor working with counterparts, and cross-cultural influencing and communication.
- The appendices discuss Army special operations forces imperatives, legal considerations, and media considerations.

Chapter 1
The Strategic Context

This chapter places security force assistance in strategic context. First, it explains security force assistance and its relationship to full spectrum operations, spectrum of conflict, and operational themes (see FM 3-0). A discussion of U.S. national strategy and defense policies that impact security force assistance follows. This chapter then examines operations and programs related to security force assistance. Finally, it discusses the relationship of security force assistance and related operations and programs.

SECURITY FORCE ASSISTANCE AND THE STRATEGIC CONTEXT

1-1. *Security force assistance* is the unified action to generate, employ, and sustain local, host-nation or regional security forces in support of a legitimate authority (FM 3-07). Security force assistance (SFA) improves the capability and capacity of host-nation or regional security organization's security forces. These forces are collectively referred to as foreign security forces. ***Foreign security forces* are forces—including but not limited to military, paramilitary, police, and intelligence forces; border police, coast guard, and customs officials; and prison guards and correctional personnel—that provide security for a host nation and its relevant population or support a regional security organization's mission.**

1-2. SFA occurs within the framework of full spectrum operations (see FM 3-0). In most situations involving this assistance, there is relatively little weight on offensive and defensive operations from a U.S. perspective. However, when U.S. forces accompany foreign security forces (FSF) in combat, the weight of offensive and defense operations will change to address the situation and align with the foreign security force's efforts. SFA is not just a stability operation, although it is a key contributor to the primary stability tasks of establish civil security and establish civil control.

1-3. SFA can be conducted across the spectrum of conflict, from stable peace to general war. This assistance could focus on improving the security forces of a host nation that is currently under no immediate threat, on paramilitary forces to counter an insurgency, or on advising FSF in major combat operations against an external threat. Figure 1-1 illustrates how SFA spans the spectrum of conflict.

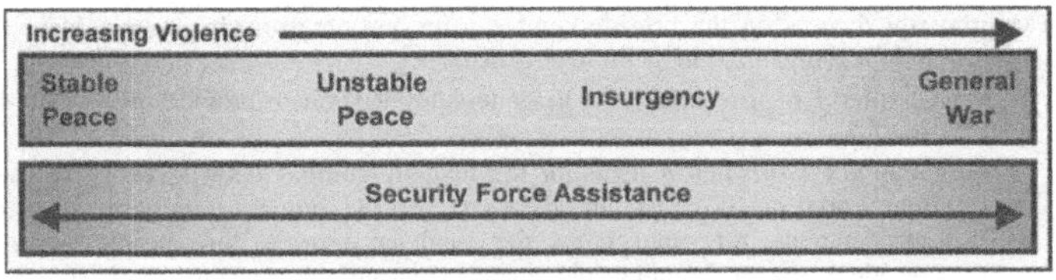

Figure 1-1. The spectrum of conflict and security force assistance

1-4. SFA can occur as part of any operational themes—peacetime limited engagement, limited intervention, peace operations, irregular warfare, and major combat operations. SFA often occurs during peacetime military engagement, peace operations, and irregular warfare. SFA involving paramilitary forces will often be limited to the irregular warfare operational theme, although paramilitary forces should eventually be absorbed, disbanded, or maintained in a long-term relationship with special operations forces.

1-5. While Army units are vital contributors, SFA is part of a comprehensive approach. This comprehensive approach should include close collaboration with military and civilian joint and multinational forces. The host-nation or regional security organization is the key actor within the comprehensive approach. It is vital that SFA is based on an objective, continuous assessment that examines the organization, training, equipment, rebuilding, and advising of the forces involved.

1-6. Law enforcement, military, intelligence, and border forces operate and cooperate within the security sector. U.S. forces understand how these units are intended to operate in the host-nation scheme, not necessarily using a U.S. model. U.S. forces should plan to help develop the respective capabilities so that these units can carry out their security functions.

1-7. Fully developed and institutionalized FSF plan, prepare, and execute at all the appropriate levels. Thus, U.S. forces must integrate SFA into the operations process at all levels. SFA should be designed to work seamlessly with the host-nation government at all levels, from ministries addressing the security sector to initial entry-level FSF training.

NATIONAL STRATEGY AND DEFENSE POLICIES

1-8. U.S. strategy and foreign policy affects all SFA. More importantly, SFA is critical to these policies as it is a primary tool for building partnership capacity with other nations. Key strategy documents for SFA include the National Security Strategy, the National Defense Strategy, the National Military Strategy, the Quadrennial Defense Review, and DODD 3000.05.

1-9. The National Security Strategy outlines the President's vision for providing enduring security for the American people in a volatile, uncertain, complex, and ambiguous strategic environment. The National Security Strategy of 2006 addresses stability operations within the broad engagement strategy for regional conflict. It supports SFA by "tailoring assistance and training of military forces to support civilian control of the military and military respect for human rights in a democratic society." This strategy also states that "[t]he world has found through bitter experience that success often depends on the early establishment of strong local institutions such as effective police forces and a functioning justice and penal system. This governance capacity is critical to establishing the rule of law and a free market economy, which provide long-term stability and prosperity."

1-10. The National Defense Strategy encompasses the Secretary of Defense's vision for the Department of Defense role in protecting the American people and U.S. interests. This strategy stems from the National Security Strategy. The 2008 National Defense Strategy highlights SFA. It states that "[o]ur strategy emphasizes building the capacities of a broad spectrum of partners as the basis for long-term security.... [l]ocal and regional conflicts in particular remain a serious and immediate problem...we will help build the internal capacities of countries at risk...[b]y helping others to police themselves and their regions, we will collectively address threats to the broader international system."

1-11. The National Military Strategy supports the National Security Strategy, National Defense Strategy, and Quadrennial Defense Review. The 2004 National Military Strategy states "[t]he Department will also work to strengthen alliances and partnerships by helping other nations increase their ability to defend themselves and protect common security interests." Working with other nations and strengthening partnerships appears throughout the National Military Strategy, which supports the importance of SFA.

1-12. The Quadrennial Defense Review is a congressionally mandated Department of Defense (DOD) review of strategy, programs, and resources. The 2006 Quadrennial Defense Review directed that joint ground forces "possess the ability to train, mentor and advise FSF and conduct counterinsurgency campaigns." This review also stated that conventional forces need to be capable of training, equipping, and advising local forces; deploying and engaging with partners; and supporting stability operations. Additionally, the review envisioned conventional ground forces continuing to perform more of the tasks traditionally performed by special operations forces. Conventional forces must understand foreign cultures and societies as well as be able to train, mentor, and advise FSF.

1-13. DODD 3000.05 provides guidance on stability operations. DODD 3000.05 establishes that stability operations are a core military mission and equal in importance to combat operations. DODD 3000.05 also reinforces the importance of SFA by stating "[t]he immediate goal often is to provide the local populace with security, restore essential services, and meet humanitarian needs. The long-term goal is to help develop host-nation capacity for securing essential services, a viable market economy, rule of law, democratic institutions, and a robust civil society."

RELATED OPERATIONS AND PROGRAMS

1-14. Several operations or programs directly or potentially relate to SFA. These operations and programs provide the context for SFA. They include security cooperation, security assistance, foreign internal defense, internal defense and development (IDAD), and security sector reform. For more information on operations related to SFA, see DOD 5105.38M, DOD security cooperation guidance, JP 3-07.1, FM 3-07, FM 3-05.137, and AR 11-31.

SECURITY COOPERATION

1-15. *Security cooperation* is all Department of Defense interactions with foreign defense establishments to build defense relationships that promote specific U.S. security interests, develop allied and friendly military capabilities for self-defense and multinational operations, and provide U.S. forces with peacetime and contingency access to a host nation (JP 3-07.1). Finally, security cooperation occurs across the spectrum of conflict and is not exclusively a peacetime activity. Security cooperation includes security assistance programs administered by the DOD as well as activities that enhance interoperability and the collective capability of combined forces (using Title 10 or exercise funding and authority). Security cooperation consists of a focused program of bilateral and multilateral defense activities conducted with foreign countries to serve mutual security interests and build defense partnerships. Security cooperation is governed by various sections of Title 10 and specific public laws addressing DOD interactions with other nations.

1-16. AR 11-31 establishes Department of the Army policy and prescribes responsibilities and procedures for the planning, integration, programming, budgeting, and execution of Army security cooperation activities. It is the policy of the U.S. Army to conduct security cooperation activities in compliance with higher-level guidance and in the execution of Army responsibilities under Title 10 and Title 22 which governs the transfer, exchange, conduct, and development of articles and services via a variety of U.S. Government programs.

1-17. Army security cooperation consists of official, cooperative, and noncombat interactions among any Army elements, Active Army, or Reserve Components. This includes any U.S. Government or nongovernment entity supporting the military and civilian joint and multinational organizations.

1-18. Commanders distinguish security cooperation from SFA based its emphasis in building relationships and capacities by using programmatic activities. Thus, security cooperation may support or be supported by SFA.

SECURITY ASSISTANCE

1-19. *Security assistance* is a group of programs authorized by the Foreign Assistance Act of 1961, as amended, and the Arms Export Control Act of 1976, as amended, or other related statutes by which the United States provides defense articles, military training, and other defense-related services by grant, loan, credit, or cash sales in furtherance of national policies and objectives (JP 3-57). Security assistance is a specific subset of security cooperation and may focus on external or internal threats.

1-20. DOD 5105.38-M describes the scope of security assistance programs in detail. Security assistance allows the transfer of military articles and services to friendly foreign governments. These transfers may be carried out via sales, grants, leases, or loans. If these transfers are essential to the security and economic well-being of allied governments and international organizations, they are equally vital to the security and economic well-being of the United States. U.S. forces can use security assistance to deliver defense weapon systems to foreign governments; to train international students; to advise other governments on improving internal defense capabilities; and to provide guidance and assistance in establishing infrastructures and economic bases for regional stability. Finally, security assistance cannot be conducted using Title 10 or exercise funds.

1-21. The military component of security assistance, implemented by DOD with policies established by the Department of State, has four principal components. They are international military education and training, foreign military sales, foreign military financing program, and peace operations. The foreign military financing program and international military education and training fall inside the military assistance budget process of the Department of State. Security assistance can also include funding peace operations.

International Military Education and Training

1-22. International military education and training contributes to internal and external security of a country by providing training to selected foreign militaries and related civilian personnel on a grant aid basis (see JP 3-07.1). These programs help to strengthen foreign militaries through training for the proper functioning of a civilian-controlled, apolitical, professional military. International military education and training serves as a foreign policy tool where the United States shapes doctrine; promotes self sufficiency in maintaining and operating U.S. acquired defense equipment; encourages the value of rule of law; and occasionally has a marked effect on the policies of the recipient governments. Foreign students, many of who occupy the middle and upper echelons of their country's military and political establishments, are taught U.S. doctrine and weapons systems employment resulting in greater cooperation and interoperability. All international military education and training activities are SFA.

Foreign Military Sales

1-23. Foreign military sales is a nonappropriated program through which foreign governments can purchase defense articles, services, and training from the United States (see JP 3-07.1). Eligible nations use this program to help build national security infrastructures. However, nations that require assistance are often unable to finance their needs. SFA supports foreign military sales when these sales include U.S. forces providing equipment and related training. It does not support foreign military sales when these sales provide equipment without related training.

Foreign Military Financing Program

1-24. The foreign military financing program provides funding to purchase defense articles and services, design and construction services, and training through foreign military sales or commercial channels (see JP 3-07.1). This program can assist nations with weak economies that would otherwise be unable to afford U.S. assistance. The foreign military financing program can fund foreign military sales. SFA and the foreign military financing program may be separate activities, although they can also be complementary.

FOREIGN INTERNAL DEFENSE

1-25. *Foreign internal defense* is the participation by civilian and military agencies of a government in any of the action programs taken by another government or other designated organization to free and protect its society from subversion, lawlessness, and insurgency (JP 3-05). U.S. foreign internal defense efforts involve all instruments of national power to support host-nation IDAD programs. The military instrument of foreign internal defense includes indirect support, direct support (not involving U.S. combat operations), and combat operations. Foreign internal defense can occur across the spectrum of conflict and within any operational theme. Foreign internal defense is often a part of irregular warfare or peacetime military engagement operational theme. SFA may support the military instrument of foreign internal defense. JP 3-07.1 discusses foreign internal defense in detail.

Note: It is a national policy decision for U.S. forces to accompany FSF on combat operations.

Military Instrument of Foreign Internal Defense

1-26. The military instrument of foreign internal defense supports the other instruments of national power through activities ranging from peacetime military engagement to major combat operations. Although foreign internal defense may include capacity building across the host nation, U.S. forces focus on combating internal threats to help the host nation maintain legitimacy and influence over the population. As such, foreign internal defense may be part of irregular warfare.

1-27. SFA and foreign internal defense support the host nation's IDAD. Consequently, forces conducting SFA must have proper mindset—they are part of a developmental and supporting effort. This applies even if SFA is part of working with a potential adversary or an existing enemy. SFA and foreign internal defense focus on working by, with, and through FSF and not conducting unilateral efforts to achieve objectives.

Indirect Support

1-28. Indirect support efforts are foreign internal defense operations that emphasize the principle of host-nation self-sufficiency. Indirect support focuses on building strong national infrastructures through economic and military capabilities that contribute to self-sufficiency. For example, indirect support could include U.S. assistance in improving a host-nation's intelligence infrastructure. Contributions from U.S. forces to indirect support come from security cooperation guidance through security assistance. They are supplemented by multinational exercises, exchange programs, and selected joint exercises. Indirect support that is focused on assisting security forces is SFA.

1-29. Although they indirectly assist those involved, military exchange programs are not SFA. Military exchange programs foster mutual understanding between forces; familiarize each force with the organization, administration, and operations of the other; and enhance cultural awareness. These exchange programs often have long-term implications for strengthening democratic ideals and respect for human rights among supported governments. However, such exchange programs (funded with Title 10 monies) do not themselves become vehicles for security assistance training, as they do not intend to assist security forces but to foster mutual understanding and enhance cultural awareness.

1-30. Joint and multinational exercises strengthen U.S. and host-nation relations and interoperability of forces, but they are not SFA. They complement security assistance and civil-military operations by validating host-nation needs and capabilities and by providing a vehicle for the conduct of humanitarian and civic assistance programs (see JP 3-07.1). Strict legal restrictions clarify the type of support that U.S. forces can provide and on the monetary limits of such support. These efforts are not SFA as they do not intend to directly assist security forces; instead, they intend to improve relations and interoperability. See appendix B.

Direct Support

1-31. Direct support does not involve combat operations but includes the use of U.S. forces providing direct assistance to the host-nation civilians or military. Direct support often focuses on civil-military operations (primarily, providing services to local civilians), psychological operations, intelligence and communications sharing, and logistic support. Sometimes direct support will be SFA, such as psychological operations, military training support, logistic support, and intelligence and communications sharing. SFA is not a civil affairs activity such as foreign humanitarian assistance, humanitarian and civic assistance, or military civic action.

1-32. Direct support differs from security assistance in that direct support is joint- or Service-funded. Direct support does not usually involve transferring arms and equipment and does not often include training local military forces. Direct support operations are normally conducted when the host nation has not attained or regained self-sufficiency and when it still faces social, economic, or military threats beyond its capability.

Combat Operations

1-33. The introduction of U.S. combat forces into foreign internal defense operations requires a Presidential decision. These operations are a temporary solution until host-nation forces can stabilize the situation and provide security for civilians. In all cases, U.S. combat operations support the host-nation IDAD program and remain strategically defensive in nature. When U.S. forces are advising, mentoring, partnering, or augmenting host-nation forces in foreign internal defense combat operations, this is SFA. When Army units conduct independent combat operations alongside FSF, this is not SFA.

1-34. U.S. tactical participation in host-nation internal conflicts requires judicious and prudent rules of engagement and guidelines for applying force. Inappropriate destruction and violence attributed to U.S. forces may easily reduce the legitimacy and sovereignty of the supported government. Adversaries may also use such incidents to fuel anti-American sentiments and assist their cause.

SECURITY RELATIONSHIPS

1-35. A complex relationship exists among security cooperation, security assistance, and the military instrument of foreign internal defense. The left side of figure 1-2 depicts this relationship, including how aspects of foreign internal defense and security assistance overlap. The right side of figure 1-2 illustrates how foreign internal defense focuses on internal threats to a host nation and how security assistance focuses on external threats. The column depicts how security (military, intelligence, and civilian), economic, and governance are considerations common to both foreign internal defense and security assistance.

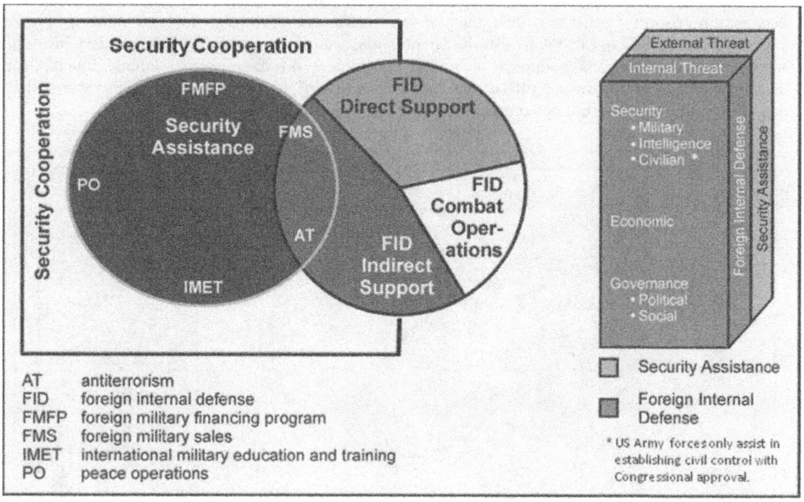

Figure 1-2. Security cooperation, security assistance, and foreign internal defense

INTERNAL DEFENSE AND DEVELOPMENT

1-36. *Internal defense and development* is the full range of measures taken by a nation to promote its growth and to protect itself from subversion, lawlessness, and insurgency. It focuses on building viable institutions (political, economic, social, and military) that respond to the needs of society (JP 3-07.1). IDAD aims to forestall or defeat the threat and to correct core grievances that prompt violence. Ideally it is a preemptive strategy. However, if an insurgency or other threat develops, it becomes an active strategy to combat that threat. To support the host nation effectively, U.S. forces, especially planners, consider the host-nation's IDAD strategy.

1-37. IDAD focuses on building viable institutions that respond to the needs of society. IDAD is a U.S. doctrinal term; it is not used universally by all countries. Internal defense and development blends four interdependent functions to prevent or counter internal threats: balanced development, security, neutralization, and mobilization. Finally, IDAD involves a cyclic interaction of execution, assessment, and adaptation. JP 3-07.1 discusses IDAD and its functions in detail.

SECURITY SECTOR REFORM

1-38. *Security sector reform* is the set of policies, plans, programs, and activities that a government undertakes to improve the way it provides safety, security, and justice (FM 3-07). Security sector reform aims to provide an effective and legitimate public service that is transparent, accountable to civil authority, and responsive to the needs of the public. It may include integrated activities to support defense and armed forces reform; civilian management and oversight; justice, police, corrections, and intelligence reform; national security planning and strategy support; border management; disarmament, demobilization, and reintegration; and concurrent reduction of armed violence. SFA is a vital component of security sector reform when security sector reform includes U.S. assistance to FSF. See chapter 6 of FM 3-07 for an additional discussion of security sector reform.

SECURITY FORCE ASSISTANCE RELATIONSHIP

1-39. Paragraph 1-35 explained and figure 1-2 supported the complex relationship among security cooperation, security assistance, and the military instrument of foreign internal defense. Figure 1-3 explains the relationship of SFA to security cooperation, security assistance, and the military instrument of foreign internal defense. SFA supports the military instrument of foreign internal defense, much of security assistance efforts, and some security cooperation efforts. The right side of figure 1-3 also shows that SFA addresses security issues, but not economic or governance issues.

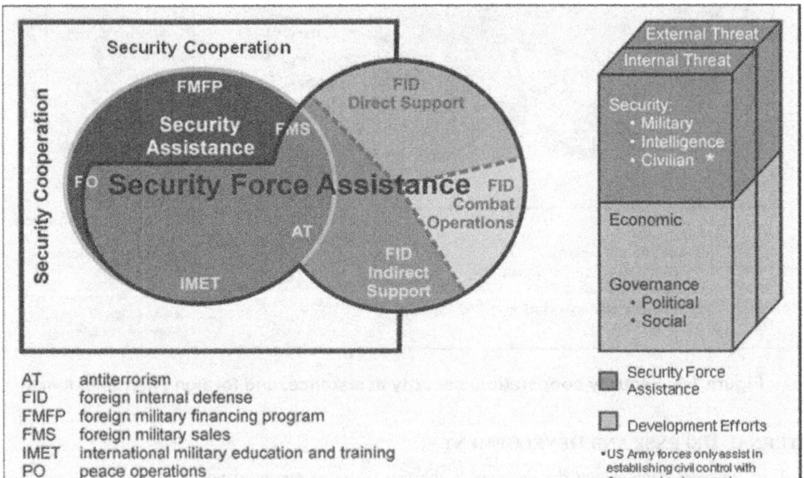

Figure 1-3. Relationship of security force assistance with security cooperation, security assistance, and foreign internal defense

Country Team

1-40. The country team will play a vital role with respect to SFA. The country team is the senior in-country U.S. coordinating and supervising body headed by the U.S. chief of mission. The chief of mission leads the country team and integrates U.S. efforts in support of the host nation. The chief of mission is the highest U.S. civil authority in a foreign country. As the senior U.S. Government official permanently assigned in the host nation, the chief of mission is responsible to the President for policy oversight of all U.S. Government programs. The Foreign Service Act assigns the chief of mission to a foreign country with responsibility for the direction, coordination, and supervision of all government executive branch employees in that country except for Service members and employees under the command of a U.S. area military commander.

1-41. The country team is composed of the senior member of each represented department or agency. As permanently established interagency organizations, country teams represent a priceless resource. They often provide deep reservoirs of local knowledge and interaction with the host-nation government and population. See FM 3-07 and JP 3-07.4 for more information on the country team.

Theater Command Structures

1-42. SFA is inherently multinational. Nations do not relinquish their national interests by participating in multinational operations. As in any multinational operation, commanders conducting SFA prepare to address issues related to legality, mission mandate, and prudence early in the planning process. In multinational operations, consensus often stems from compromise.

1-43. Political considerations heavily influence the ultimate shape of the command structure in which U.S. forces will conduct SFA. However, participating nations should strive to achieve unity of command for the operation to the maximum extent possible. All participants must understand missions, tasks, responsibilities, and authorities. While well defined in U.S. doctrine, command relationships are not necessarily part of the doctrinal lexicon of nations with which the United States may operate in coalition or combined operations.

1-44. Command and control relationships of higher echelons take precedence when conducting SFA. As a result, the brigade combat team (BCT) may be a supported command from a U.S. perspective as well as acting as a supporting command for FSF in the BCT's area of operations. For example, a BCT may support a joint interagency task force conducting international police training. Another example is a joint command supporting a BCT with joint fires; intelligence, surveillance, and reconnaissance; or other capabilities while the BCT supports FSF in training. The basic structures for multinational operations fall into one of three types: lead nation, integrated, or parallel command. Alliance and coalition commands use these structures.

Lead Nation Command Structure

1-45. A lead nation command structure exists when all member nations place their forces under the control of one nation. The lead nation command has a dominant lead nation command and staff arrangement with subordinate elements retaining strict national integrity. This command may also have an integrated staff and multinational subordinate forces. Integrating the staff allows the commander to draw upon the expertise of allied or coalition partners in areas where the lead nation may have less experience. This is the most desirable command structure for SFA.

Integrated Command Structure

1-46. Multinational commands organized under an integrated command structure provide unity of effort in a multinational setting. An integrated, combined command has several distinguishing features. Besides a designated single commander, the staff consists of representatives from all member nations. Subordinate commands and staffs are integrated into the lowest echelon necessary. Although effective for SFA, the integrated command structure works less effectively than the lead nation command structure.

Parallel Command Structure

1-47. Under a parallel command structure, no single force commander is designated. The coalition leadership develops a means for coordination among the participants to attain unity of effort. They can accomplish this by using coordination centers. Nonetheless, because of the absence of a single commander, the use of a parallel command structure should be avoided if at all possible. This is the least desirable command structure for SFA.

Alliance Command Structures

1-48. In combined commands, national political objectives are addressed and generally subsumed within multinational force objectives at the alliance treaty level. Combined command relationships often reflect either an integrated command structure or a lead nation command structure. In combined operations such as SFA, these structures should be used to the maximum practical extent. Combined command and force structures often mirror the degree of allied member participation. Senior military officers from member nations often lead subordinate commands. Effective operations within an alliance, including SFA, require senior political and military authorities to agree on the type of command relationships that will govern the operations of the forces. Potential political sensitivities associated with SFA can impact command relationships and operating procedures.

Coalition Command Structures

1-49. Coalitions often form in response to unforeseen crises that occur outside the area of an established alliance, which is often the case for SFA. Coalition command relationships routinely evolve as a coalition develops. Coalitions often have one of two basic structures: lead nation or parallel. For SFA, all participating coalition members should augment the host-nation headquarters staff with representatives. These representatives might include designated deputies or assistant commanders, planners, and logisticians. Such staffing provides the coalition commander with representative leadership and accessible expertise on the capabilities of the coalition members as well as facilitates the planning process. This integration of the multinational command elements into the coalition planning process should occur as early as practical.

Integration of Conventional and Special Operations Forces

1-50. Many criteria determine when conventional forces, special operations forces, or a combination are appropriate to conduct SFA. Both force levels and force characteristics suggest optimal, acceptable, and undesirable force package options in planning and resourcing SFA. Options for the deployment of a modular brigade augmented for SFA, a select number of conventional military transition teams, or special operations forces depend on conditions of the operational environment, priorities of the IDAD strategy, overall U.S. national policy, and forces available.

1-51. Rarely will U.S. forces conducting SFA be homogenous in terms of conventional or special operations forces, but rather contain both. FM 3-0 and other Army doctrine define command and control relationships and support relationships when both forces operate in the same area. Recent operational experience has shown that, in improving the effectiveness of FSF, the supported commander conducting SFA may be either a conventional force or special operations force commander. Subordinate commanders and staffs of both conventional and special operations forces may act in the supporting role to the U.S. ambassador. Typically this is done through the security assistance officer or the defense attaché at the American Embassy.

1-52. If U.S. forces are to train a large number of FSF in a short time, then the requirement may exceed the capacity of available special operations forces. Large SFA missions occur at the geographic command level or higher headquarters. Planners at these levels determine the number of conventional and special operations forces as part of their mission analysis aimed at training enough FSF in the given time frame. To best prepare for a large SFA, U.S. forces might deploy one or more augmented modular brigades supported by Army special operations forces. These forces conduct complementary foreign internal defense tasks under Title 22 authorities. Another option involves using special operations forces training FSF over several years while using a broad mix of funding and authorities. Another option uses a modular brigade to support a select number of company- or battalion-sized foreign units.

1-53. A force analysis of how best to integrate conventional and special operations forces to accomplish SFA considers host-nation and U.S. national policies. Skill sets among conventional and special operations forces trainers, advisors, and mentors avoid universally excluding or including final, integrated force packages for SFA. Planners leverage skill sets by integrating and coordinating conventional and special operations units with assigned tasks within SFA. Force analysis of the FSF to be trained often suggests a two-phase approach to training. This analysis also suggests a transition to the FSF by either conventional or special operations forces using basic and advanced training. The type of training depends on the nature of the foreign security force being trained. Additionally, integrating the advisor may be a phased process using either conventional or special operations forces as the initial advisor. This method lends itself to the classic crawl-walk-run approach to assisting FSF.

Chapter 2
Framework

This chapter discusses a doctrinal framework for the conduct of security force assistance. This security force assistance framework is composed of the mindset required of units and Soldiers, imperatives for success, inherent tasks and activities, and the three types of security force assistance.

MINDSET

2-1. Conducting successful security force assistance (SFA) requires a specific mindset. This mindset focuses on working by, with, and through foreign security forces (FSF) to support the host nation's internal defense and development (which includes local security requirements) or regional organization's charter. Soldiers conducting SFA must also understand that legitimacy is vital. The relevant population must perceive FSF as legitimate for long-term success. Those conducting SFA must understand that the military instrument of national power is only one part of a comprehensive approach. The imperatives of SFA provide the foundation for proper mindset.

Note: Host nation includes partner organization.

2-2. SFA focuses on how FSF organizations, institutions, systems, capabilities, and limitations—not U.S. organizations, systems, procedures, and equipment—can be enabled to address the host-nation or partner organization problems. Local forces have advantages over outsiders. They inherently understand the local culture and behavior that outsiders simply lack. To tap into those advantages, advisors must resist blatant military solutions. To overcome the temptation to do what they know and do best, whether relevant or not to the situation, advisors must accept that they are bound by unique situations.

IMPERATIVES

2-3. The imperatives of SFA come from the historical record and recent experience. These imperatives do not replace the principles of war or the principles of joint operations. Rather, they provide focus on how to successfully conduct SFA. The six imperatives apply to SFA at every level of war, for any echelon, and for any Soldier.

2-4. Since there is a close relationship between SFA and Army special forces conducting foreign internal defense, SFA planners should also consider the twelve Army special operations forces imperatives, especially at the tactical level (see appendix A).

UNDERSTAND THE OPERATIONAL ENVIRONMENT

2-5. An in-depth understanding of the operational environment—including available FSF, opposing threats, and civil considerations—is critical to planning and conducting effective SFA. Units and Soldiers conducting SFA must clearly understand the theater, population, and FSF with which they are working, especially FSF capabilities. Diplomatic, informational, military, economic, sociological, psychological, and geographic research and understanding are essential prerequisites for successful SFA. Tactically, successful SFA requires identifying the friendly and hostile decisionmakers, their objectives and strategies, and the ways they interact. Further, the conditions of the operational environment can rapidly change, and those conducting SFA must anticipate these changes and exploit potential and possibly fleeting opportunities.

PROVIDE EFFECTIVE LEADERSHIP

2-6. Leadership, a critical aspect of any application of combat power, proves important in the dynamic and complex environments associated with SFA. The operational environment in which SFA occurs places a premium on effective leadership at all levels, from the most junior to the most senior general officer and agency director. Leading FSF or a combined group of U.S. and foreign security forces is inherently more challenging due to differences in culture, language, training, and other aspects. Leaders focus on transitions; their decisions move the foreign security force toward independent operations.

BUILD LEGITIMACY

2-7. Ultimately, SFA aims to develop security forces that contribute to the legitimate governance of the local populace. Significant policy and legal considerations may apply to SFA activities. Legitimacy is the most crucial factor in developing and maintaining internal and international support. The United States cannot sustain long-term SFA without legitimacy. Legitimacy is a concept that goes beyond a strict legal definition; it includes the moral and political legitimacy of a host-nation government or partner organization. Local civilians and the international community determine the government's legitimacy based on collective perception of the credibility of its cause and methods.

MANAGE INFORMATION

2-8. Successful SFA disseminates timely and protected relevant information, integrates it during planning, and leverages it appropriately during execution. Effective and efficient information management supports decisionmaking throughout capacity building. Managing information encompasses the collection, analysis, management, application, and preparation of information. Maintaining continuity between rotating SFA elements is critical. For long-term SFA efforts, FSF expect more professional abilities from U.S. advisors. To meet expectations and success in general, units conducting SFA establish and integrate lessons learned. Capturing and incorporating evolving tactics, techniques, and procedures and aspects of the operational environment is vital; with each successive rotation, planners strive to maintain continuity and momentum. Finally, planners synchronize effective information engagement with the entire mission. See appendix C. SFA operations inherently affect the dynamic operational environment.

ENSURE UNITY OF EFFORT

2-9. SFA often includes many actors, making unity of effort essential for success. SFA will include U.S. and foreign security forces, including conventional forces, special forces, or a combination. Other civilian and military joint and military organizations are often involved in SFA. Planners integrate them into one cohesive effort.

SUSTAIN THE EFFORT

2-10. Sustainability consists of two major components: the ability to sustain SFA effort throughout the operation and the ability of the FSF to sustain their operations independently. While each situation will vary, Army personnel conducting SFA must avoid assisting FSF in techniques and procedures beyond the FSF's capability to sustain. U.S. tactics, techniques, and procedures must be modified to fit the culture, educational level, and technological capability of the FSF. Those involved in SFA must recognize the need for programs that are durable, consistent, and sustainable by both the U.S. and FSF. They must not begin programs that are beyond the economic, technological, or cultural capabilities of the host nation to maintain without U.S. assistance. Such efforts are counterproductive.

TASKS

2-11. SFA aims to establish conditions that support the partner's end state, which includes legitimate, credible, competent, capable, committed, and confident security forces. This requires a force capable of securing borders, protecting the population, holding individuals accountable for criminal activities, regulating the behavior of individuals or groups that pose a security risk, and setting conditions in the operational area that enable the success of other actors.

2-12. This section addresses the six SFA tasks: organize, train, equip, rebuild and build, and advise and assist. When supporting the development of FSF, commanders understand their role with the host-nation or regional security organization. These tasks facilitate SFA planners to assess and allocate resources based on conditions.

ORGANIZE

2-13. Organize is a SFA task that encompasses all measures taken to assist FSF in improving its organizational structure, processes, institutions, and infrastructure. U.S. forces must understand the existing security organizations of FSF to better assist them. Subsequently, SFA personnel may help the host nation organize its security forces to meet the needs of its security environment.

2-14. Organizing a foreign security force depends on the host nation's social and economic conditions, cultural and historical factors, and security threats. SFA aims to create an efficient organization with a command, intelligence, logistic, and operations structure viable for the host nation. Conventional forces with limited special purpose teams (such as explosive ordnance disposal) are preferred. Doctrine or standing operating procedures should apply across the force, as should unit structures. The organization must facilitate the collection, processing, and dissemination of intelligence across and throughout all security forces. As the foreign security force strengthens, U.S. leaders and trainers should expect more independent organizational decisions.

2-15. The host nation determines the structure of its military forces, to include approving all organizational designs. These may include changing the numbers of forces, types of units, and internal organizational designs.

2-16. Organization should address all FSF, from the ministerial level to the patrolling police officer and soldier. Building a competent FSF infrastructure—including civilian information systems—is critical for success. Commanders work with ministries responsible for national and internal security, including the ministry of national defense, the interior ministry, and the justice ministry. Commanders assess strengths and weaknesses of the ministerial organization as well as training requirements of their employees. The joint and multinational advisor team at the ministry level helps the host nation develop a procurement and management system to meet its requirements.

2-17. Organizing a foreign security force requires resolving issues related to—

- Recruiting.
- Promotion screening and selection.
- Pay and benefits.
- Leader recruiting and selection.
- Personnel accountability.
- Demobilization of security force personnel.

Recruiting

2-18. Recruiting is critical when establishing FSF. The recruiting program assimilates local culture and themes that resonate with the populace. The program ensures that FSF have members from all major demographic groups. Joint and multinational partners should encourage and support efforts to recruit from minority populations. Mobile recruiting can target specific areas, ethnic groups, or tribes. Forces should encourage moderate groups and factions within hostile or potentially hostile ethnic groups to join the foreign security force. Often host-nation governments resist recruiting disaffected ethnic groups. However, even moderate success in recruiting from these groups provides enormous payoffs. It builds the security forces' legitimacy and often quiets legitimate fears of such groups regarding their relationship to the host-nation government. Effectively disarming, demobilizing, and reintegrating former enemies or other armed groups should be part of the overall plan. These programs should be mentioned in recruiting efforts.

2-19. A proper recruiting program specifies appropriate behavioral, physical, and moral standards. Ideally, recruits are centrally screened and inducted. Recruiting centers reside in safe areas safe. All recruits undergo a basic security check and are vetted against lists of suspected enemies. Foreign agencies and personnel perform this screening. Membership in illegal organizations needs to be carefully monitored. Past membership need not preclude a person from joining the foreign security force; however, any ongoing relationship with an illegal organization requires constant monitoring. FSF personnel limits the members of an illegal unit, tribal militia, or other militant faction to a single military or police unit.

Promotion Screening and Selection

2-20. Selection for promotion must stem from proven performance and aptitude for increased responsibility. Objective evaluations ensure promotion is by merit and not through influence or family ties. One method of selection identifies the most competent performers, trains them, and recommends them for promotion. The second method identifies those with social or professional status within the training group, then trains and recommends them for promotion. The first method may lead to more competent leaders but could be resisted for cultural reasons. The second method ensures the new leader will be accepted culturally but may sacrifice competence. The most effective solution comes from combining the two methods.

Pay and Benefits

2-21. Appropriate compensation precludes a culture of corruption in FSF. Spending the money needed for adequate wages and producing quality security forces is less costly than ending up with corrupt and abusive forces that alienate the populace. Paying the police adequately is especially important; their duties and contact with civilians often expose them to opportunities for corruption.

2-22. Payroll procedures and systems are also vitally important. They must be transparent and accountable, so FSF members receive their full pay and entitlements. Centralized administration of compensation through secure, automated systems minimizes the risk for corruption and manipulation; however, the supporting infrastructure will have to be available or developed over time.

2-23. Effective FSF can help improve host-nation social and economic development through the benefits each member receives. Every recruit should receive a basic education, job training, and morals and values inculcation.

Leader Recruiting and Selection

2-24. Officer candidate standards should be high. Candidates should have good health and pass an academic test with higher standards than the test for enlisted recruits. Recruiters carefully vet officer candidates to ensure that they do not have close ties to any radical or enemy organization.

2-25. Noncommissioned officers should be selected from the best enlisted FSF members. Objective standards, including proficiency tests, should be established and enforced to ensure that promotion to their ranks comes from merit. Many armies lack a professional noncommissioned officer corps; establishing one for a foreign security force may prove difficult. In the meantime, adjustments will have to be made, placing more responsibility on commissioned officers.

Personnel Accountability

2-26. Host-nation leaders carefully track and account for FSF. Proper personnel accountability reduces corruption, particularly in countries with manual banking systems where soldiers are paid in cash. In addition, many personnel failing to report for duty can indicate possible attacks, low unit morale, or enemy and militia influences on the foreign security force.

Demobilization of Foreign Security Force Personnel

2-27. Host nations develop programs to keep a class of impoverished and disgruntled former officers and soldiers from forming. As the foreign security force mature, officers who perform poorly or fail to meet the standards will need to be removed. Providing some form of government-provided education grants or low-interest business loans enables discharged personnel to earn a living outside the military. All soldiers who serve for several years and are then removed should receive a lump-sum payment or pension to ease their transition to civilian life. These programs should not apply to those guilty of major human rights abuses or corruption. Transition or demobilization planning should start as soon as commanders anticipate the need (it may not be required in all cases). Similar programs may be required when demobilizing nongovernment militias.

TRAIN

2-28. Train is a SFA task to assist FSF by developing programs and institutions to train and educate. These efforts must fit the nature and requirements of their security environment. The imperatives of Army special operations forces for facilitating a comprehensive approach and engaging the threat discriminately apply to the train task (see appendix A).

Training the U.S. Trainers

2-29. Soldiers assigned training missions should receive training on the requirements of developing FSF. The training should emphasize the host nation's cultural background, introduce its language, and provide cultural tips for developing a good rapport. Training should also include protection for troops working with FSF. U.S. trainees must become familiar with FSF organization and equipment, especially weapons not in the U.S. inventory. This training must emphasize the following:

- Sustaining training and reinforcing individual and team skills.
- Using the smallest possible student-to-instructor ratio.
- Developing host-nation trainers.
- Training to standards—not to time.
- Providing immediate feedback; using after action reviews.
- Respecting the culture but learning to distinguish between cultural practices and excuses.
- Learning the language.
- Working with interpreters.

2-30. U.S. forces should show respect for local religions and traditions. They should willingly accept many aspects of the local and national culture, including food (if sanitation standards permit). U.S. forces must stress that they do not intend to undermine or change the local religion or traditions. However, they have to strive to reduce how dysfunctional social practices affect the foreign security force's ability to conduct operations. U.S. trainers and advisors must have enough awareness to identify and stop inappropriate behavior, or at least report it to the appropriate chains of command.

Establishing Training Standards

2-31. Training in SFA involves many of the individual and collective skills performed in conventional military operations. All levels of training for all components should include values training. Metrics for evaluating units should include subjective measures, such as loyalty to the host-nation government, as well as competence in military tasks. Soldiers know how to evaluate military training. However, the acceptance of values, such as ethnic equality or the rejection of corruption, may be a better measure of training effectiveness. Gauging this acceptance is far more difficult than evaluating task performance. While the operational environment varies widely, FSF and trainers can still establish clear measures to evaluate the training of individuals, leaders, and units.

2-32. Effective training programs require clear, detailed individual, leader, and unit performance standards. These standards take into account cultural factors that directly affect the ability of the individual or unit to operate. For example, training a staff or unit to conduct effective operations requires more time in countries where the average soldier is illiterate. Building a security force from the ground up takes far more time than creating one around a trained cadre. Thus, using existing military personnel to form units and cadres for units often proves better than creating novice security forces.

2-33. Often, poorly trained leaders and units commit more human rights violations than well-trained, well-led units. Leaders and units unprepared for the pressure of active operations tend to use indiscriminate force, target civilians, and abuse prisoners. These actions can threaten the popular support and government legitimacy essential for SFA success. Badly disciplined and poorly led FSF can facilitate insurgent recruiting and propaganda efforts.

2-34. Setting realistic measures for FSF and following through on training plans consume time. The pressure is strong to find training shortcuts, employ quick fixes, or train personnel on the job. Trainers should resist such approaches. Such approaches often create more problems than they solve. However, trainers should also avoid the temptation to create long, complex training programs based on unrealistic standards. Effective programs account for the host nation's culture, resources, and short-term security needs. No firm rules exist on how long particular training programs should take. Trainers can use existing and historical training programs to determine how long training should take. To a certain extent, the enemy threat dictates the length of training. As security improves, training programs can expand to facilitate achievement of the long-term end state.

Training Foreign Security Forces

2-35. Members of FSF develop through a systematic training program (individual training and education as well as collective training). The program builds basic skills, then teaches them to work as a team, and finally allows them to function as a unit. Basic military, intelligence, or law enforcement training focuses first on basic skills such as legal considerations, first aid, marksmanship, and fire discipline. Leaders have training in tactics, including patrolling, urban operations, and legal evidence collection. Everyone must master rules of engagement and the law of armed conflict. FSF units should train to standard for conducting the major operations they will face.

Training Leaders

2-36. The effectiveness of FSF directly relates to the quality of their leadership. Building effective leaders requires a comprehensive program of officer, staff, and specialized training. The ultimate success of any SFA effort depends on creating viable FSF leaders able to carry on the fight at all levels and build their nation on their own.

2-37. The standards of leader training reinforce different levels of authority within the foreign security force. Clearly established responsibilities for commissioned and noncommissioned officers specify what is expected of recruits and leaders. To ensure civilian control, subordinate relationships to civilian authorities also are reinforced. In addition, training should establish team dynamics. In some cultures, security forces may need training to understand the vital role of members not in primary leadership positions. For example, noncommissioned officers may be a new or different concept for some FSF.

Employing Newly Trained Forces

2-38. SFA must build the morale and confidence of FSF. Committing poorly trained and badly led forces results in high casualties and invites tactical defeats. While defeat in a small operation may have little strategic consequence in a conventional war, often a small tactical defeat of FSF has serious strategic consequences. If the foreign security force fails, the local populace may begin to lose confidence in the host-nation government's ability to protect them. FSF must prepare for operations so they have every possible advantage. As much as possible, FSF should begin with simpler missions. As their confidence and competence grows, these forces can assume more complex assignments. Collaborating with joint or multinational units can help new forces to adjust to combat stress.

2-39. Newly trained units should enter their first combat operation in support of more experienced foreign units. Operational performance of such inexperienced organizations should be carefully monitored and evaluated to correct weaknesses quickly. The employment plan for FSF should allow enough time for additional training after each operation. Gradually introducing units into combat allows the command to identify poor leaders for retraining or other actions. Competent leaders are also identified and given greater authority and responsibility.

EQUIP

2-40. Equip is a SFA task encompassing all efforts to assess and assist FSF with the procurement, fielding, and sustainment of equipment. All equipment must fit the nature of the operational environment. The SFA principle of ensuring long-term sustainment is a vital consideration for the equip task.

2-41. The strategic plan for security force development should outline equipment requirements. Appropriate equipment is affordable and suitable against the threat. Forces must be able to train on the equipment. Interoperability may be desired in some cases. A central consideration includes the host nation's long-term ability to support and maintain the equipment.

2-42. The initial FSF development plan should use a comprehensive approach with local, regional, national, and other appropriate actors. Fundamentally, it should support the internal defense and development strategy. It should include phases with objectives for FSF to meet over three to four years. Since potential enemies adapt rapidly and situations change, commanders must continually assess the direction and progress of developing FSF.

2-43. The requirement to provide equipment may be as simple as assisting with maintenance of existing formations or as extensive as providing everything from shoes to communications and investigation kits. If the enemy uses heavy machine guns and rocket-propelled grenades, then FSF need comparable or better equipment. This especially applies to police forces, which are often lightly armed and vulnerable to well-armed enemies.

2-44. Primary considerations should include maintainability, ease of operation, and long-term sustainment costs. Few developing nations can support highly complex equipment. In most operations, having many versatile vehicles that require simple maintenance is often better than having a few highly capable armored vehicles or combat systems that require extensive maintenance. Effective maintenance systems for FSF often begin with major maintenance performed by contractors. The program then progresses to arrangements with U.S. forces as they train foreign personnel to conduct the support mission.

2-45. Sources of materials for FSF include U.S. foreign military sales (not technically SFA), multinational or third-nation resale of property, contracts with internal suppliers, or purchases on the international market. The organizations responsible for equipping FSF require flexibility to obtain equipment that meets their needs for quality, timeliness, and cost. As part of their training, FSF also need to learn property accountability to reduce corruption and ensure proper equipment usage. FSF provide equipment the same level of control and protection that U.S. forces provide for similar equipment. (See AR 12-1 and DOD 5105.38M.)

REBUILD AND BUILD

2-46. Rebuild and build is a SFA task to assess, rebuild, and build the existing capabilities and capacities of FSF and their supporting infrastructure. This task requires an in-depth analysis of the capability, capacity, and structures required to meet the desired end state and operational environment. Some FSF may require assistance in building and rebuilding, while other FSF may only need assistance in building.

2-47. FSF need infrastructure support. People need buildings for storage, training, and shelter. Often requirements include barracks, ranges, motor pools, and other military facilities. Construction takes time; the host nation needs to invest early in such facilities if they are to be available when needed. Forces must plan to protect any infrastructures, including headquarters facilities, since these building are attractive targets for insurgents. (See FM 3-34.400 for information on hardening measures to increase infrastructure survivability and improve protection.)

2-48. During some operations, such as counterinsurgency, foreign military and police forces often operate from local bases. Building training centers and unit garrisons requires a long-term force-basing plan. If possible, garrisons should include housing for the host-nation soldiers and their families; government-provided healthcare for the families; and other attractive benefits.

2-49. The host nation may need to make large investments in time and resources to restore or create the nationwide infrastructure necessary to effectively command and control FSF. The host nation will build functional regional and national headquarters and ministries as well as local bases and police stations.

ADVISE AND ASSIST

2-50. Advise and assist is a SFA task in which U.S. personnel work with FSF to improve their capability and capacity. Advising establishes a personal and a professional relationship where trust and confidence define how well the advisor will be able to influence the foreign security force. Assisting is providing the required supporting or sustaining capabilities so FSF can meet objectives and the end state. The level of advice and assistance is based on conditions and should continue until FSF can establish required systems or until conditions no longer require it. Leaders cannot permit FSF to fail critically at a point that would undermine the overall effort. See chapter 6 for a discussion on advisors.

ACTIVITIES

2-51. SFA is inherently a developmental effort. Success is measured by the increase in the foreign security force's capability, capacity, competency, commitment, and confidence in areas in which U.S. forces are assisting them and as a whole. SFA activities include plan and resource, generate, employ, transition, and sustain. Figure 2-1 depicts the SFA activities. These activities blend over time based on the capability and capacity of FSF.

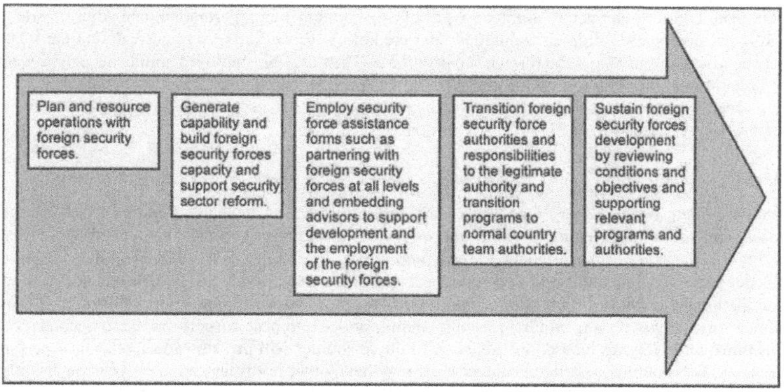

Figure 2-1. Security force assistance activities

PLAN AND RESOURCE

2-52. The plan and resource SFA activity begins as commanders understand the operational environment and determine the requirements of FSF. It also ensures that the United States provides SFA that achieves the objectives and end state of both the host nation and United States. The host nation and United States should then analyze the resource requirements and efforts so that developing FSF have sufficient and appropriate resources. Chapter 4 provides additional detail on planning for facilities and infrastructure.

GENERATE

2-53. The generate SFA activity includes generation of FSF, including the forces, leaders, and warfighting function capabilities (movement and maneuver, intelligence, fires, sustainment, command and control, and protection) based on the desired end state.

EMPLOY

2-54. The employ SFA activity involves FSF transitioning from force generation to mission employment. It does not rely on the maturity of the force or supporting institutions as a whole but is focused on the foreign element in question. Conditions determine when to use an element of FSF. Leaders of U.S. and foreign security forces assess the risk associated with employment and mitigate that risk as much as possible.

TRANSITION

2-55. The transition SFA activity defines the transition between two security forces, when applicable. This could be a transition of authority between U.S. forces to a new foreign security force. Another example could be from a regional foreign security force to a foreign security force with U.S. advisors. It could also be from a host-nation military force to a host-nation police force, with or without U.S. advisors.

SUSTAIN

2-56. The sustain SFA activity occurs when the institutional capacity of the foreign security force has been developed to a point where it is self sustaining. It may continue to have SFA contact through combined exercises, educational opportunity exchange, intelligence sharing, and foreign military sales.

TYPES

2-57. The U.S. or other actors use three types of SFA—augmenting, partnering, and advising—to develop confident, competent, capable, committed, and credible FSF. These types of SFA define the relationship between outside actors and the foreign security force. They may be employed simultaneously, sequentially, or in combination. The progression and types of SFA are determined by the operational environment, the assessment of the FSF, and by resources available. Each of these types requires decidedly different requirements, objectives, and legal considerations.

ADVISING

2-58. Advising is the primary type of security force assistance. Advising is the use of influence to teach, coach, and advise (see chapter 6) while working by, with, and through FSF. This type of SFA relies on the ability of the advisor to provide relevant and timely advice to FSF. Advising helps FSF conduct independent decisionmaking and operations. Advisors may also provide FSF with direct access to joint and multinational capabilities such as air support, artillery, medical evacuation, and intelligence. However, the advisor and advisor team require proper manning and equipment to perform these secondary support functions while staying focused on advising. Multiple sources will pressure advisors as they perform their missions. For example, nearby maneuver units may have other priorities and may focus on accomplishing their missions with or without FSF.

2-59. Advisor teams require a clearly defined and structured chain of command under which to operate. This is not only for logistics and support, but also keeps the advisor focused on developing FSF. Advisor teams will often find themselves answering to their higher military assistance group, the brigade combat team they are attached to, and the host-nation unit with which they are embedded. A well-defined chain of command alleviates confusion concerning who tasks or monitors the team's progress as well as ensures constant and adequate team sustainment.

2-60. Advisors are not partners; U.S. forces act as partners. Advising and partnering are complementary but inherently different activities. Advising requires relationship building and candid discourse to influence development of a professional security force. Partnering incorporates training with combined operations to achieve the same SFA goals. Advisors perform partnership shaping functions, shape discussions with their counterparts, and create opportunities for the partner units.

PARTNERING

2-61. Partnering attaches units at various levels to leverage the strengths of both U.S. and foreign security forces. As a foreign security force's capabilities mature, the echelon and degree of partnering decrease. As the foreign security force conducts more autonomous operations, U.S. forces still provide quick reaction forces and other assistance as appropriate.

2-62. Partner units should establish combined cells for intelligence, operations, planning, and sustainment. While effective coordination is always required and initial efforts may require completely fused efforts, FSF should eventually build the capability and capacity to conduct all efforts autonomously. These combined cells have several functions. They support transparent operations and a comprehensive approach. They also enhance the relationships among U.S. and foreign security forces by demonstrating trust. Finally, they develop the capacity of FSF in key staff areas. However, combined cells are not without risk. Operations and plans always at risk of compromise and prudent precautions should be taken.

2-63. Another partnering technique establishes relationships among command and staff elements of U.S. and foreign security forces. It requires less reorganization. It also allows counterparts to understand each other since foreign personnel can observe U.S. personnel performing their duties. This construct works well when the foreign culture is extremely sensitive to maintaining honor. In this case, foreign personnel can observe without having to expose their lack of knowledge or appear subordinate.

2-64. Unit partnerships do not replace advisor roles or functions. If partnering and advising are used in combination, it forms a three-part relationship amongst FSF, advisors, and the partner units. Partner units should look to the advisor to identify, shape, and facilitate operational partnering opportunities and training events. Advisors support U.S., coalition, and partner unit objectives, but, depending on the operational phase, the partner unit may support advisors or advisors may support the partner unit. Therefore, some level of advisor skills training should be included in the partner unit training program if those units will be conducting SFA activities.

AUGMENTING

2-65. Augmenting is an arrangement where FSF provide individuals or elements to combine with U.S. units, or U.S. individuals or elements combine with FSF. Augmentation improves the interdependence and interoperability of U.S. and foreign security forces. Augmentation can occur at many levels and in many different forms. For example, a U.S. squad can be augmented with host-nation individuals, a U.S. company can be augmented with a host-nation platoon, or a U.S. battalion can be augmented with a company from a foreign security force. Similarly, augmentation can be of short duration for a specific operation or of a longer duration for an enduring mission. Augmenting immerses FSF in a U.S. environment to provide language and cultural awareness to the U.S. unit. U.S. forces can also augment FSF.

COMBINING TASKS, ACTIVITIES, AND ASSESSMENT

2-66. Execution of successful SFA links its activities, tasks, and assessment. Figure 2-2 depicts how to arrange SFA activities, tasks, and assessment in time and purpose. The tasks that occur in one SFA activity will normally continue throughout operations. For example, organize, train, equip, rebuild and build, and advise and assist tasks will continue throughout the SFA activities of employment, transition, and sustain.

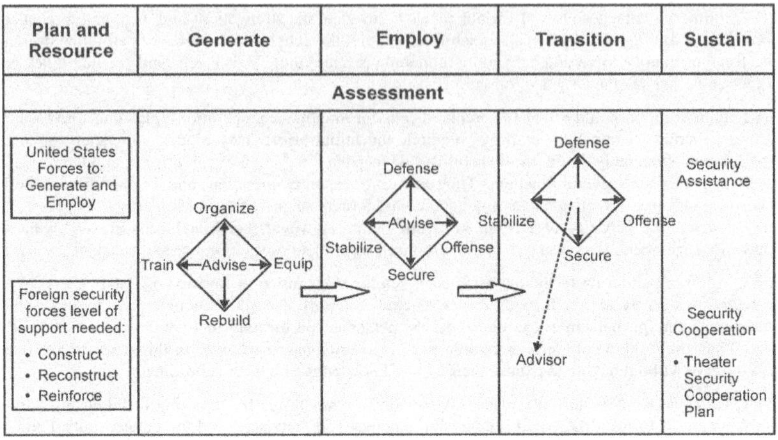

Figure 2-2. Security force assistance tasks, activities, and assessment

Plan and Resource

2-67. The plan and resource SFA activity determines the level of support and number of U.S. forces needed. As all efforts in SFA this activity is part of a comprehensive approach. Planners assess whether they will be constructing a foreign security force from the ground up, reconstructing a foreign security force based on existing capabilities and structure, or merely reinforcing an existing foreign security force. Understanding what capability and capacity the foreign security force should have and what capability and capacity they currently have is vital. Commanders and planners can then determine the size and structure, resources, and funding for the U.S. unit or units that will facilitate or assist the generation of FSF. Chapters 3 and 4 discuss planning and execution considerations for SFA.

Generate

2-68. In the generate SFA activity, U.S. forces focus on organizing, training, equipping, and advising FSF. During generation of FSF, the condition of the existing foreign security force's infrastructure and capabilities determine the assistance needed from U.S. forces. The generate activity adapts to the needs and requirements of the foreign security force. It can be designed to take the individual recruit from initial entry through to an employable unit, or it can augment an existing foreign unit with additional or enhanced capabilities. Like all SFA efforts, generate focuses on bringing a foreign security force's capabilities to a level where it can conduct its mission. With help from advisors in the next phase, FSF can work toward independent and sustained employment. In large SFA efforts, an organization may be required to command and control this activity.

2-69. Once elements of FSF are trained, the focus shifts from organizing, training, equipping, and rebuilding to employing the foreign security force. However, not all FSF units and organizations are ready for employment at the same time. As one unit is employed, others may be part of the generation activity, while still others may be completing the transition activity. FSF may be employed for offensive, defensive, or stability operations while other FSF may be employed against internal or external threats. The primary SFA task for U.S. forces is to advise and assist to facilitate the foreign security force operations.

Transition

2-70. The transition activity focuses on removing the U.S. elements and advisors once FSF are ready to conduct independent operations. All involved actors work in a comprehensive approach to assess when this can be achieved and to coordinate the transition itself. As FSF gradually progress toward the transfer of authority, close relationships forged between the FSF and their partners prove essential to sustainable development and successful transition. Genuine relationships engender trust and confidence, enabling increased responsibility and a well-executed transition process. These relationships also foster a clear understanding of command responsibilities and authorities. Such an understanding ensures FSF approach transition prepared to assume the full weight of their future role in the security sector. Success in developing FSF often depends more on relationships and personalities at the unit level than any other factor.

Chapter 3
Brigade Operations Process

This chapter examines operational considerations for the modular brigade augmented for security force assistance (SFA). It uses battle command and the operations process to examine these considerations.

BATTLE COMMAND

3-1. *Battle command* is the art and science of understanding, visualizing, describing, directing, leading, and assessing forces to impose the commander's will on a hostile, thinking, and adaptive enemy. Battle command applies leadership to translate decisions into actions—by synchronizing forces and warfighting functions in time, space, and purpose—to accomplish missions (FM 3-0). Commanders are central to battle command for brigade combat team (BCT) operations just as they are for any other operation. However, there are key considerations for successful battle command of a BCT conducting SFA. Most importantly, BCT commanders must work with their foreign counterparts, which may require facilitating operations when the foreign security forces (FSF) commander has the lead for the operations process.

UNDERSTAND

3-2. Successful SFA requires that BCT commanders thoroughly understand the operational environment. Conditions of the operational environment affect what SFA commanders provide. Commanders must understand the conditions that make up the current situation, including the relevant population, other actors, and FSF. From this understanding, commanders visualize desired conditions that represent a desired end state. Commanders must understand the capacity and capabilities of foreign security. The host-nation or regional security organization helps to determine desired conditions associated with the foreign security forces. After envisioning a desired end state, commanders then conceptualize how to change the current situation to the desired end state.

3-3. BCT commanders must understand the host-nation's security and political concerns. Security concerns focus on internal and external threats, their capability to affect the operational environment, and the foreign security force's strength and capability to counter those threats. Politically, the considerations associated with the security sector are the legitimacy of the foreign security force, FSF capability and capacity, host-nation or partner organization policy, and the ability to support the host-nation and U.S. Government objectives.

3-4. Commanders use the information obtained from intelligence preparation of the battlefield and intelligence, surveillance, and reconnaissance (ISR) to determine objectives and requirements for assessing and mitigating risk. Risk applies to how well FSF, U.S. Government, host-nation, partner organization, and international organizations can tolerate changes in the operational environment, as well as the challenges and conditions inherent to the operation.

VISUALIZE

3-5. *Commander's visualization* is the mental process of developing situational understanding, determining a desired end state, and envisioning the broad sequence of events by which the force will achieve that end state (FM 3-0). Visualization begins with assignment of the mission, which provides the focus for developing the commander's visualization. However, visualization must be continuous as SFA is dynamic. During planning, commander's visualization provides the basis for developing plans and orders.

3-6. The nature and extent of assistance provided to FSF are determined by this desired end state. SFA is often a slow process that does not lend itself to quick solutions. To ensure long-term success, commanders and host-nation or regional security organizations clarify early what conditions they desire. These conditions include the desired end state for FSF, which often include being—

- **Flexible**. FSF can accomplish broad missions. Flexibility requires an effective command and control organization appropriate for the partner or host-nation. This can include both internal and external threats.
- **Proficient**. The level of proficiency of FSF should include—
  - The capability of working effectively within a comprehensive approach.
  - The capability to integrate their operations with those of multinational partners.
  - Nonmilitary security forces competent in maintaining civil order, enforcing laws, controlling borders, securing key infrastructure, and detaining criminal suspects.
  - Nonmilitary security forces thoroughly trained in modern police ethos and procedures and which understand the basics of investigation, evidence collection, and proper court and legal procedures.
- **Self-sustained**. FSF can manage their own equipment throughout its lifecycle and perform administrative support.
- **Well led**. Leaders at all levels possess sound professional standards and appropriate military values; they are selected and promoted based on competence and merit.
- **Professional**. FSF are—
  - Honest, impartial, and committed to protecting and serving the entire population; they operate under the rule of law and respect human rights.
  - Loyal to the partner organization or host-nation government and serving partner or national interests; they recognize their role as the servants and not masters.
- **Integrated into society**. In the case of a host nation, FSF represent the host nation's major social groups; they are not seen as elements of just one faction.

3-7. These conditions may vary in different countries, but well-trained FSF should—

- Provide reasonable levels of security from external threats without threatening regional security.
- Provide reasonable levels of internal security without infringing on the populace's civil liberties or posing a coup threat to the host-nation government.
- Be founded upon the rule of law, including international principles of the rule of law.
- Be sustainable after joint and multinational forces depart.

3-8. When addressing internal threats, foreign military and police forces often perform unconventional functions. The military may have to fill an internal security role usually reserved for the police, or police may have forces so heavily armed that they would normally be part of the military. In the near term, FSF should—

- Integrate military capabilities with those of local, regional, and national police.
- Maintain the flexibility to transition to more conventional roles of external and internal defense based on long-term requirements.

3-9. To achieve this end state and intermediate objectives, the host nation should develop an internal defense and development (IDAD) strategy (with multinational assistance if required). This strategy should have a section on IDAD that addresses all aspects of SFA.

3-10. U.S. doctrine divides force development into domains: doctrine, organization, training, materiel, leadership and education, personnel, and facilities (DOTMLPF), which may be useful to analyze the IDAD. DOTMLPF elements are tightly linked, simultaneously pursued, and difficult to prioritize. Commanders monitor progress in all domains. Often military forces involved in such programs are tempted to impose their own doctrine and judgment on the host nation. Foreign security force doctrine, like the remaining DOTMLPF domains, must be appropriate to host-nation capabilities and requirements. Paragraphs 3-59 through 3-70 use DOTMLPF as a tool to assess FSF.

DESCRIBE

3-11. After BCT commanders visualize how to conduct SFA, they describe it to their staffs and subordinates to facilitate shared understanding of the mission and commander's intent. Commanders ensure subordinates understand the visualization well enough to begin planning. Commanders express their initial visualization as—

- Initial commander's intent.
- Planning guidance, including an initial concept of operations.
- Commander's critical information requirements.
- Essential elements of friendly information that must be protected.

Commander's Intent

3-12. Like any operation, SFA requires a clear, concise statement of what the force must do and the conditions the force must meet to succeed with respect to the enemy, terrain, and desired end state. In such an operation, the end state should include a description of desired conditions with respect to FSF capability and capacity. The commander's intent links the mission and concept of operations. It describes the key tasks and end state that, along with the mission, form the basis for subordinates' initiative.

3-13. A sample end state could discuss—

- How civilians perceive FSF as credible and legitimate and representing a legitimate authority.
- How FSF have the capability to support stability and security, retain the initiative and freedom of movement, disrupt enemy activities, influence the population, operate with other friendly forces, and sustain themselves.

3-14. Like any operation, how the BCT supports the higher organization's mission is vital. Additionally, the BCT supports FSF. Some sample key tasks are:

- Support the organization of a division of foreign infantry and two battalions of foreign police.
- Train leaders of FSF to create a cadre that can sustain their own training programs.
- Oversee the equipping of FSF and provide new equipment training.
- Support the rebuilding or building of infrastructure required to develop FSF.
- Advise FSF to develop the required capabilities and competencies associated with shared objectives.
- Assist the partner's security forces in meeting their operational requirements.

Planning Guidance

3-15. Commanders develop planning guidance for the staff from the commander's visualization. Planning guidance may be as broad or detailed as circumstances require; however, it must convey to the staff the essence of the commander's visualization. Due to the complex and dynamic nature of FSF, SFA often requires detailed planning guidance. Commanders ensure the staff understands their visualization while still permitting the staff to explore different options. Some considerations are:

- Friendly courses of action (COAs) with differing SFA forms and differing methods on how to augment, partner, or advise.
- Command and control architecture, intelligence architecture, and communications architecture to support dispersed SFA.
- Enemy COAs as they seek to defeat friendly efforts with respect to SFA activities.
- Enemy adaptation given the development of the foreign security force.
- Guidance for security during the conduct of SFA, including the BCT securing its advisor teams. The BCT determines the staging, needs, and ad hoc arrangements of advisor teams.
- ISR guidance to include identifying the focus of effort as it relates to SFA.

Commander's Critical Information Requirements

3-16. Commander's critical information requirements (CCIRs) will be developed for SFA just as for any other operation. These requirements are what commanders need to make decisions or the information they think is important for understanding the operational environment. These CCIRs should be compatible with the FSF commander's information requirements. CCIRs include priority intelligence requirements and friendly force information requirements.

3-17. Priority intelligence requirements are vital to support decisionmaking. During SFA, priority intelligence requirements must be focused on the enemy and the operational environment, not on FSF. U.S. forces do not focus intelligence efforts, such as collection, on the foreign forces they assist.

3-18. Friendly force information requirements provide the information that commanders need to understand the conditions of their own forces, as well as the conditions of the foreign security forces. For example, a BCT commander may direct the staff to better define a foreign security force's critical equipment shortages or maintenance requirements. During execution, commanders may focus on when FSF meet certain objectives or conditions. This information allows the commander and FSF commander to make adjustments, such as changes to the task organization, support requirements, or even the type of SFA.

Essential Elements of Friendly Information

3-19. An *essential element of friendly information* is a critical aspect of a friendly operation that, if known by the enemy, would subsequently compromise, lead to failure, or limit success of the operation, and therefore should be protected from enemy detection (FM 3-0). Essential elements of friendly information should be determined through a comprehensive approach. Information sharing and trust are vital elements of SFA. U.S. commanders must be aware of the inherent risks in this context and employ prudent risk mitigation efforts. Thus, successful SFA requires a balance between sharing information and risk mitigation efforts. Although essential elements of friendly information are not CCIRs, they have the same priority.

Intelligence, Surveillance, and Reconnaissance

3-20. *Intelligence, surveillance, and reconnaissance* is an activity that synchronizes and integrates the planning and operation of sensors, assets, and processing, exploitation, and dissemination systems in direct support of current and future operations. This is an integrated intelligence and operations function. For Army forces, this activity is a combined arms operation that focuses on priority intelligence requirements while answering the commander's critical information requirements (FM 3-0). For the brigade, these efforts will often support FSF. Moreover, the BCT must have means that allow effective intelligence collaboration, such as an intelligence fusion cell with the participation of all appropriate actors.

3-21. ISR efforts focus on the enemy and operational environment within the context of SFA. ISR efforts in SFA are not used on the foreign security force. Rather, U.S. forces assess the foreign security force within a comprehensive approach.

DIRECT

3-22. BCT commanders direct all aspects of operations, although a comprehensive approach, including FSF and other appropriate actors is fundamental to successful SFA. The commander's direction takes different forms during planning, preparation, and execution. Commanders make decisions and direct actions based on their situational understanding, which they maintain by continuous assessment and coordination with FSF. SFA is directed by—

- Preparing and approving plans and orders with FSF.
- Assigning and adjusting missions, tasks, task organization, and control.
- Positioning units to maximize their support to FSF, anticipate threat actions, or create or preserve options.
- Positioning key leaders to ensure observation and supervision of FSF at critical times and places.

- Adjusting SFA and other support priorities and allocating resources based on FSF priorities, as well as opportunities and threats.
- Changing support arrangements, either internally or with respect to FSF.

ASSESS AND LEAD

3-23. Effective battle command requires commanders to continuously assess and lead. Assessment helps commanders to better understand current conditions and determine how the operation is progressing and how best to support FSF. Based on an assessment that is the product of a comprehensive approach, commanders modify plans and orders to better accomplish the mission. This may include changing the type or combination of SFA, support, sustainment, or task organization. If their assessment reveals a significant variance from the original shared commander's visualization, a comprehensive approach must be used to reframe the visualization and develop a new plan.

3-24. Commanders in a BCT conducting SFA lead by example and presence, as Army commanders do in any operation. However, commanders have the additional challenge of leading by, with, and through FSF. Command posts for BCTs conducting SFA have similar challenges in that they should be integrated headquarters as much as practical.

OPERATIONS PROCESS

3-25. As they would in any other mission, BCT commanders conducting SFA balance their time and the staff's time and resources among four major activities in a continuous learning and adaptive cycle called the operations process. The *operations process* consists of the major command and control activities performed during operations: planning, preparing, executing, and continuously assessing the operation. The commander drives the operations process (FM 3-0). While simple in concept, the operations process is complex in execution. The operations process activities may be sequential or simultaneous.

3-26. Figure 3-1 depicts the operations process activities as a cycle that is driven by battle command. This cyclic process applies equally to a BCT conducting major combat operational theme or operations within an irregular warfare operational theme. During the operations process involving SFA, commanders consider the presence and inclusion of other actors, including FSF. Policy, the operational environment, and the forms of SFA determine how U.S. forces, FSF, and other actors interact in the operations process.

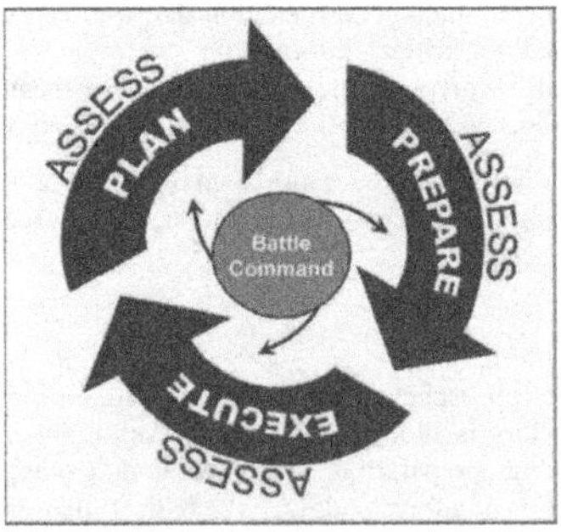

Figure 3-1. The operations process

PLAN

3-27. Planning for SFA, like any other operation, begins either with the anticipation of a new mission or the receipt of mission as part of the military decisionmaking process. *Planning* is the process by which commanders (and the staff, if available) translate the commander's visualization into a specific course of action for preparation and execution, focusing on the expected results (FM 3-0). SFA requires a comprehensive approach to ensure the U.S. commander's visualization synchronizes with policy and the foreign commander's visualization. SFA is the major requirement for a BCT conducting SFA, but the mission may require the brigade to perform other tasks.

3-28. Continuous and open to change, planning for SFA includes identifying how to best assist the FSF and developing a sequence of actions to change the situation. This planning must occur within a comprehensive approach that includes FSF, U.S. forces, and other actors. Similarly, this plan must be nested within policy, IDAD strategy, the campaign plan, and any other higher-echelon plans. It also involves anticipating consequences of actions and developing ways to mitigate them.

3-29. The plan, which includes the commander's intent, provides a common understanding to U.S. and foreign security forces on the actions to take. Plans and orders also provide decision points and branches that anticipate options that enable the force to adapt as the operation unfolds. This is especially important for SFA, as these operations tend to be prolonged efforts. BCTs conducting SFA often rotate before achieving all objectives. As a result, the planning team should establish objectives and milestones that can be achieved during the brigade's mission. These BCT objectives and milestones must support higher-echelon plans, including the campaign plan and IDAD strategy.

Functional Requirements

3-30. The composition of advisor teams is subject to objectives and conditions. For example, security conditions may require three armored vehicles, a crew-served weapon, and nine personnel to conduct independent movement. Another example involves the difference in the number of advisors. Newly generated FSF may require many advisors whereas experienced FSF may only require assistance in a few warfighting functions and consequently need fewer advisors. Understanding the functional requirements provides flexibility to the units assisting FSF. Three examples of determining functional requirements based on the understanding the operational environment include the BCT, a partnering or augmenting unit, and an advisor team.

Brigade Combat Team

3-31. A BCT headquarters has specific functions and requirements. These can include, but are not limited to—

- The functions of—
  - Carrying out SFA tasks and assisting FSF in support of objectives and the end state.
  - Facilitating partnerships between U.S. and foreign security forces.
  - Requesting, coordinating, and synchronizing resources.
  - Providing command and control of subordinate units and assigned advisor teams.
  - Coordinating with other commands and adjacent units.
  - Coordinating with other participants integrated within the comprehensive approach.
  - Developing, implementing, and synchronizing information engagement in assigned areas.
  - Developing relationships and an understanding with appropriate actors involved in SFA.

3-32. The requirements of—

- Having enough staff to handle the communications traffic generated by conditions.
- Being able to analyze information to aid commander and subordinate elements.
- Being able to articulate resource requirements in support of subordinate elements and efforts to conduct SFA tasks.
- Being organized and equipped to conduct operations and support and sustain FSF.
- Including regionally oriented host-nation officers to provide regional expertise.

Partnering or Augmenting Unit

3-33. Functions and requirements depend on the degree and level of partnering or augmenting. No prescriptive way to partner or augment exists; it is based on specific conditions and subject to the needs of policy, higher headquarters guidance, and needs of the FSF. A BCT may designate one battalion to partner or augment while designating other battalions to advise. A partnering or augmenting unit conducting SFA has specific functions and requirements. These can include, but are not limited to—

- The functions of—
  - Partnering with foreign security under the BCT level.
  - Increasing capability, capacity, competency, confidence, and commitment of FSF by conducting combined tactical operations.
  - Providing feedback on performance of FSF and certain shortfalls to embedded advisors and BCT headquarters.
  - Conducting sustainment and medical training with FSF at their home station or on operations.
  - Reporting conditions in the operational environment.
- The requirements of—
  - Having U.S. forces that are capable, competent, committed, and confident.
  - Being interoperable with FSF.
  - Being available for an appropriate period.

Military Advisor Team

3-34. The functions and requirements of a military advisor team are based on conditions—requirements, availability, and footprint. A military advisor team conducting SFA has specific functions and requirements. These can include, but are not limited to—

- The functions of—
  - Under the BCT headquarters, advising, training, and assisting FSF to which it is assigned.
  - Increasing capability, capacity, competency, confidence, and commitment of FSF by providing advice and support during battalion level and higher operations.
  - Assessing partner leaders, staff, and certain shortfalls to BCT headquarters.
  - Conducting sustainment training with FSF at their home or on operations.
  - Reporting on conditions in the operational environment.
- The requirements of—
  - Being manned with trained personnel who are properly equipped so as to be capable of executing their mission in current conditions.
  - Providing personnel with rank, education, and experience corresponding to the echelon and type of unit being advised.

Support and Sustainment Requirements

3-35. Support and sustainment requirements play a vital role in COA development (see chapter 6). The BCT may operate across a broad geographical area when conducting or supporting SFA. Partnering may be on certain levels with FSF and may support numerous embedded advisor teams or other actors. Since FSF often have varied relationships with higher echelons and institutions, the BCT may be involved vertically and spatially. These considerations may require more support and sustainment requirements than an initial analysis might indicate. Support and sustainment considerations include, but are not limited to—

- Distribution of FSF to which U.S. forces are providing SFA.
- The capability of FSF to provide support, sustainment, and medical functions.
- The technical capability and interoperability of U.S. and foreign security forces.
- The infrastructure that supports SFA.
- Overall security conditions where SFA occurs.

- The available contracting support.
- The ability of the environment to accept and absorb the U.S. forces' footprint.

Determining a Course of Action

3-36. COAs for SFA require an understanding of the breadth and width of the security sector. They may have to account for the roles ranging from the individual soldier or policeman to the ministerial level. COA development examines the roles other actors, for example, a gendarme or paramilitary force, in the security sector. Because the United States does not have a gendarme, U.S. forces may not fully appreciate its importance.

3-37. COAs for SFA also need to be flexible and inclusive. Commanders consider the following when examining a partner's security sector:

- The status and challenges of the host-nation security sector or regional security organization.
- The composition and disposition of the partner's security sector.
- Supporting institutions.
- Links among ministries, institutions, and components of the security sector.
- Partner capability and capacity shortfalls and impediments to their development.
- Environmental requirements to enable the building of security capability and capacity.
- Functional requirements to assist building capability and capacity.
- The impact of other actors within the comprehensive approach.

3-38. Once planners analyze the security sector, they can develop COAs that best combine SFA tasks, activities, and types (see chapter 2). While the advising type of SFA can be of any duration, partnering and augmenting tend to be enduring if used.

3-39. All SFA types tend to be more effective if they are long-term efforts. Advising fosters a more personal relationship and infers a stronger commitment to see the partner succeed. Partnering and augmenting include U.S. and foreign security forces conducting operations together. There are levels of partnering. Different advisory relationships use advisor teams embedded within various partner echelons and units, use one advisor team to advise multiple entities, or have a senior leader visit a partner unit to give advice.

3-40. Using any form or combination of forms depends on conditions (such as the level of competency), capability of FSF, and resources available. Policy, guidance, or formal agreements can also help commanders determine which form to use. Historically, a combination of forms appears to be the most effective. However, combining forms often proves more resource intensive and affected by time, geography, and the number of forces. COAs should remain flexible with respect to the form of SFA. Some example changes in conditions are the increased capability of one element in the security sector over another, the desire to increase one element over another based on partner objectives, or the increased targeting of one element over another by the enemy.

3-41. If a desired COA is not feasible given the scope of the security sector or due to other considerations, it may be useful to sequence SFA forms with portions of FSF. For example, a COA that affects smaller portions of the FSF, but in the right area and at the right time, may do more for achieving an objective or end state than a COA that attempts to support all FSF at one time.

3-42. BCTs are task-organized and augmented to accomplish a specific mission. How BCTs are specifically task-organized and augmented is based on the phase and types of SFA that policy and higher headquarters have directed. Chapter 4 describes augmentation in detail.

3-43. Attached units or units under operational control are task-organized under different headquarters as necessary to accomplish assigned and implied SFA missions. Deciding to task-organize specific formations for SFA may reflect previously approved contingency plans or result from current planning efforts. As these missions end, units return to their parent headquarters or are further task-organized by the controlling headquarters.

3-44. As with other operations, gaining commanders task-organize for SFA by designating operational control, attached, or support relationships to another unit or headquarters. Within each BCT, the brigade commander task-organizes all organic, assigned, attached, or operational control units to best suit the operational environment and types of SFA. Commanders may choose to use a support relationship (direct support, reinforcing, or general support reinforcing) among units of the BCT instead of a command relationship. This can also apply to relationships with foreign units.

3-45. SFA is often highly dispersed, as BCTs may have individuals or units assisting FSF over a wide area. This dispersion increases communications and sustainment requirements for the BCT. The nature of SFA also requires BCTs to help FSF meet requests for assets. To do so, the BCT and other units may need to task-organize to provide tactical support or sustainment. Dispersion and the nature of SFA often necessitate a decentralized leadership approach to facilitate effective and rapid support of FSF.

3-46. BCT commanders use the three types of SFA—advising, partnering, or augmenting—to accomplish the mission. The exact nature of a unit's assigned mission, the operational variables, and the mission variables drive modifications to headquarters and maneuver units. For example, a commander may need a unit tasked with advising, a unit tasked with partnering, and a unit tasked with augmenting. In this case, the unit tasked to advise would provide teams for the appropriate FSF while other foreign units continue their traditional missions. The unit that is partnering may provide the host-nation staff to assist the FSF headquarters. The third unit may provide squads to augment foreign platoons. Units may request additional support, sustainment, and medical support. However, BCTs conducting SFA must retain a reserve to conduct full spectrum operations.

3-47. Based on the overall mission analysis, including an analysis of functions and requirements, commanders and staffs begin to determine how to task organize. Due to the level of detail that may be involved, task organization should begin early in the military decisionmaking process, especially during COA development. No standard task organization exists since conditions vary. For example, a squad of infantry and three additional leaders may work as an advisor team for one mission, where another mission may require an advisor team with specialists for different warfighting functions. The art in SFA involves determining where, when, and if to allocate resources; it involves either breaking up or simply moving personnel and units from one foreign unit to another.

3-48. The techniques used by BCT commanders and staffs to allocate other resources and support are just as valid in SFA as in other types of operations. For example, if it is determined that the most important effort is advising, then the BCT commander may weight that effort with additional resources, such as communications equipment or experienced personnel.

3-49. BCTs emphasize unity of effort with FSF. Ideally, interoperability and shared resources extend to the lowest level. This means having one tactical operations center, dining facility, clinic, motor pool, and so on. Not only does this use resources more efficiently, but it also facilitates understanding, trust, and rapport between U.S. and foreign security forces. For Soldiers to effectively conduct SFA work by, with, and through FSF requires a close rapport and high level of trust.

3-50. Part of COA development for SFA includes an analysis of where to focus the brigade's subordinate units. This analysis includes determining which subordinate units support key elements of FSF. The brigade's subordinate units may support provincial reconstruction teams or transition teams that are not organic to the brigade. Another method is for the brigade may provide organic forces to make up transition teams or the military portion of provincial reconstruction teams or transition teams. Transition teams may include, but are not limited to police, border, point of entry, or military transition teams. Figure 3-2 depicts an example for a brigade providing forces as part of a provincial reconstruction team as well as providing the other transition teams. Figure 3-2 is an example of how a brigade may focus its subordinate units. Additionally, the teams are only examples. All of these teams may not be necessary for all brigades, and some brigades may require other types of advisor teams.

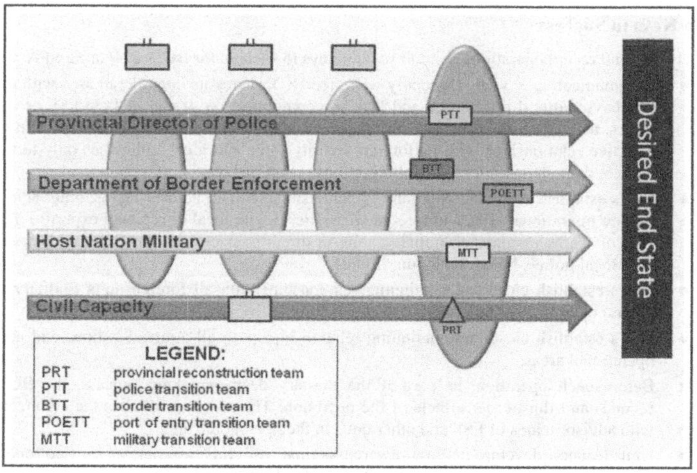

Figure 3-2. Example of brigade focus for subordinate units

PREPARE

3-51. *Preparation* consists of activities performed by units to improve their ability to execute an operation. Preparation includes, but is not limited to, plan refinement; rehearsals; intelligence, surveillance, and reconnaissance; coordination; inspections; and movement (FM 3-0). Like any operation, preparation for SFA creates conditions that improve friendly forces' opportunities for success. It facilitates and sustains transitions, including those to branches and sequels, which are of vital importance for the often dynamic operational environment for SFA. Chapter 5 covers preparation for units conducting SFA, including the BCT.

EXECUTE

3-52. *Execution* is putting a plan into action by applying combat power to accomplish the mission and using situational understanding to assess progress and make execution and adjustment decisions (FM 3-0). Execution of SFA focuses on supporting the needs of the FSF and, like many other operations, is often characterized by phases and transitions between phases. Chapter 4 examines BCT phases for SFA.

Acceptance of Prudent Risk

3-53. SFA, like any other operation, requires prudent, disciplined risk-taking focused on winning rather than preventing defeat. Because uncertainty exists in all operations, every decision involves risk. Among key elements of the art of command are deciding how much risk to accept and minimizing the effects of accepted risk (see FM 6-0). Due to SFA's inherent collaborative nature, determining prudent risk is complex and requires a comprehensive approach.

3-54. Risk reduction measures identified in planning add to the plan's flexibility during execution. A flexible plan can mitigate risk by partially compensating for a lack of intelligence. Again, SFA requires a thorough comprehensive approach to analyzing and agreeing upon risk reduction measures.

Execution Keys to Success

3-55. History and recent operations point to several keys to success for BCTs executing SFA:

- Communication is vital, especially when the BCT shares its operational area with other entities that have cultural differences and lack of communication. It can include FSF or multinational forces, as well as U.S. forces with another mission. For example, an advisor team can develop effective relationships with its foreign security force and local authorities only to have another unit conduct an operation that unhinges those relationships.
- BCTs establish close and continuing relationships with all advisor teams, other actors operating in their operational area, and foreign area officers with local or regional expertise. In addition to advisor teams assigned to military units in direct contact with the BCT, advisors may embed with local police, border, and other forces.
- BCTs establish close and continuing relationships with all foreign units (military, police, and others) operating in their operational areas.
- BCTs establish close and continuing relationships with all political entities and actors in their operational areas.
- Before each operation, as part of the military decisionmaking process, the BCT considers second- and third-order effects of the operation. The team coordinates the commander's intent with advisor teams of FSF and other units in the operational area.
- Unit leaders develop cultural awareness and use this awareness so that operations and relationships achieve the desired end state. In short, the closer a maneuver commander works with advisor teams and the more they interact with local political and cultural leaders, the better the overall chances of mission success.

ASSESS

3-56. The operations process and effective battle command require commanders to continuously assess and lead. Assessment helps commanders to better understand current situations and how SFA is progressing. *Assessment* is the continuous monitoring and evaluation of the current situation, particularly the enemy, and progress of an operation (FM 3-0). Assessment is conducted during planning, preparation, and execution and is evaluated against measures of effectiveness and performances to make decisions and adjustments (see FM 6-0).

3-57. Assessment of SFA requires a comprehensive approach that analyzes the overall operational environment and focuses on FSF and their impact. SFA assessment is performed in close collaboration with the U.S. country team, the host-nation government, and other multinational partners. These partners continually assess the security situation and its influence on other operations. From the assessment, planners develop short-, mid-, and long-range plans. As circumstances change, so do the plans. An insurgency might require employing host-nation forces at earlier stages of development. Some existing security forces may be so dysfunctional or corrupt that the organizations ideally should be disbanded rather than rehabilitated. In some cases, commanders will need to be replaced before their units will become functional. Political realities, however, may make both prospects undesirable. Minimizing and marginalizing particularly corrupt units or leaders while focusing on other units can be effective in this case.

3-58. While every situation differs, BCT leaders assess the following factors throughout planning, preparation, and execution of the operation:

- The operational environment.
- The civil considerations of the operational environment (see appendix B of FM 6-0).
- The core grievances and the threat (possibly including the prerequisites of insurgency).
- FSF elements.
- Methods, successes, and failures of FSF security efforts.
- State of training at all levels, and the specialties and education of leaders.

- Equipment and priority placed on maintenance.
- Logistic and support structure and their ability to meet the force's requirements.
- Extent of acceptance of ethnic and religious minorities.
- Laws and regulations governing the security forces and their relationships to national leaders.

Initial Assessment

3-59. An initial comprehensive assessment offers the BCT a baseline for establishing the scope of effort required. This assessment includes a troop-to-task analysis that determines the type and size of forces needed. FSF may require complete reestablishment or only assistance to increase capacity. They may completely lack a capability (such as internal affairs, corrections, or formal schools for leaders) or they may require temporary reinforcement. As with other military operations, efforts to assist FSF should reinforce success, not failure. For example, instead of building new police stations in every town, improve the good stations and use them as models for other organizations.

3-60. BCT leaders need to make decisions on what shortfalls to address first. The extent of the threats combined with resource limitations inevitably forces commanders to set priorities. However, making efforts to legitimize the host-nation leaders by encouraging them to make the decisions often proves more successful.

3-61. A key effort for SFA is the initial assessment of the DOTMLPF and policy capabilities of a foreign security force. During the early phases of a campaign, a geographic combatant command receives an initial assessment or a unit operating in the area has an assessment of FSF. The incoming BCT updates this assessment as it begins SFA.

3-62. Assessments require objective and subjective reporting, which may require a classification level. At a minimum, a BCT should assess—

- Doctrine.
- Organization.
- Training.
- Materiel and equipment.
- Leadership.
- Personnel.
- Command, control, communications, and intelligence.
- Operational effectiveness.

3-63. The doctrine assessment provides the BCT with an analysis on how the FSF operate. This assessment could focus on formal doctrine or more localized practices.

3-64. The organization portion of the overall assessment provides a perspective on how the FSF are organized. This includes the array of foreign institutions and units.

3-65. The level of training for FSF is a vital consideration. While this assessment is partly subjective, it should objectively examine individual and collective training strategies and mission-essential task lists. In the absence of mission-essential task lists, the BCT should compare FSF to U.S. equivalents.

3-66. The materiel and equipment assessment includes, but is not limited to, weapons, vehicles, equipment, and communications. It should be divided into equipment on hand, equipment readiness, and sustainment. Equipment on hand is quantitative and based on the type of equipment available to security operations. Equipment readiness reflects the operational readiness condition or serviceability of on-hand equipment. Sustainment readiness provides a qualitative and quantitative assessment that reflects how well FSF perform supply, maintenance, transportation, and medical functions. This includes facilities, systems, and structures.

3-67. The leadership assessment analyzes FSF commanders, officers, noncommissioned officers, and other leaders. This includes an assessment of staff members.

3-68. The foreign personnel assessment is quantitative and examines manning, staff manning, and administrative capabilities. It also examines officers, noncommissioned officers, and other personnel.

3-69. The command, control, communications, and intelligence assessment is qualitative and should reflect how well FSF conduct these functions.

3-70. The operational effectiveness assessment examines the foreign security force's ability to execute operations. This portion of the assessment should not be confused with the examination of training, as relatively high levels of training do not necessarily mean that FSF will perform in operations. This assessment is highly subjective.

Assessment During Operations

3-71. Assessment for the BCT conducting SFA should be part of a comprehensive approach. Assessment is directly tied to the commander's decisions throughout planning, preparation, and execution. Critical to effective assessment is developing criteria to evaluate progress toward task accomplishment, achievement of objectives, and attainment of the end state conditions. Assessment criteria are expressed as measures of effectiveness or measures of performance.

3-72. During operations, BCT and FSF commanders and staffs periodically assess the underlying framework of the plan itself. This involves reexamining the logic and assumptions used to develop the original plan, and it requires a deliberately conceived effort to learn as a part of the operation. BCT and FSF collaboration among higher, lower, and adjacent commanders, staffs, and other actors, backed up by qualitative and quantitative assessments, contributes to this learning. Based on this reexamination, commanders update their visualization, direct changes to the order, or develop an entirely new plan to adapt the force to better accomplish the mission.

Chapter 4

The Modular Brigade Augmented for Security Force Assistance

This chapter examines the typical phases for the brigade combat team conducting security force assistance, including a discussion of baseline augmentation by phase. The final section examines some of the subordinate units and elements of the modular brigade augmented for security force assistance.

THE BRIGADE COMBAT TEAM

4-1. The brigade combat team (BCT) is the cornerstone for Army modularity and is designed to operate at the tactical level across the spectrum of conflict. The modular brigade can be augmented, based on the requirements of the operational environment, with enabling assets and capabilities to support distributed security force assistance (SFA). However, the BCT augmented for SFA retains the capability to conduct full spectrum operations—offense, defense, and stability. Any of the three modular BCTs—heavy, infantry, or Stryker—can support SFA. Battalion-sized maneuver, fires, reconnaissance, and sustainment subordinate elements of the BCT support foreign security forces (FSF).

4-2. The gaining theater army force tailors all modular Army forces for use in major combat operations and contingencies. The theater army commander, working for the geographic combatant commander, determines the mix of forces and capabilities required for the BCT. This force tailoring can be done as part of the joint deliberate planning process or as a result of crisis action planning. Force tailoring includes providing BCTs with additional forces, personnel, or capabilities, including the possibility of an embedded provincial reconstruction team. Additional assets and capabilities can include, but are not limited to, command and control, communications, sustainment, engineering, military police, and intelligence.

4-3. In recent operations, the BCT has proven its versatility and capability to learn at the individual and collective levels. It supported other U.S. units conducting SFA and directly increased the effectiveness of FSF. The BCT has operated under varying levels of enemy activity, host-nation capabilities, objectives, and other conditions.

4-4. Because the BCT can operate in nonpermissive and permissive environments, it can conduct SFA across the spectrum of conflict. Thus, it can support civilian and military joint and multinational actors. The support can range from movement security and sustainment to augmentation of reconstruction teams and security for elections.

4-5. BCTs conducting SFA may support FSF development, assist FSF operations, and support and assist the development of host-nation institutions and infrastructure. While providing this support to the host nation, the BCT remains capable of conducting full spectrum operations independently.

BRIGADE COMBAT TEAM PHASES

4-6. If a force lacks the means to accomplish its mission in a single effort, then commanders normally phase the operation. A *phase* is a planning and execution tool used to divide an operation in duration or activity. A change in phase usually involves a change of mission, task organization, or rules of engagement. Phasing helps in planning and controlling and may be indicated by time, distance, terrain, or an event (FM 3-0). Phasing SFA assists in planning and controlling these dispersed and complicated operations. These phases often involve key changes in supporting SFA tasks and types employed.

4-7. BCT phases for SFA are initial response, transformation, and fostering sustainability. These phases mirror the stability operations framework described in FM 3-07 and are based on the operational environment. SFA can start in any phase or may even move to a previous phase due to changes in the conditions of the operational environment.

4-8. Figure 4-1 illustrates an SFA operation that includes all three SFA phases. The horizontal axis depicts an improving security situation, and the vertical axis depicts security capability. The example depicts an ideal situation, although this may not be the case in actual execution. Issues that arise require proper assessment and potentially the use of branches or sequels.

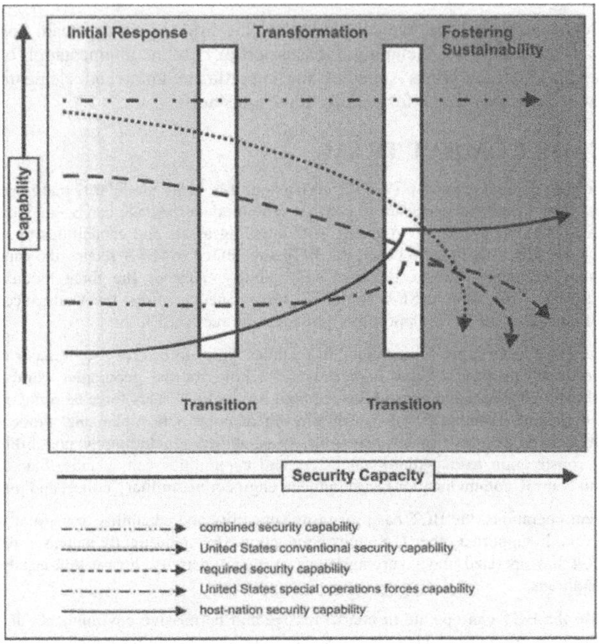

Figure 4-1. Example of phases of security force assistance

4-9. In figure 4-1, the combined security capability represents the overall security capacity of all security forces involved; it is the sum of U.S. conventional security, host-nation security, and the U.S. special operations forces security capabilities. This example assumes that the combined security capability exceeds the required security capability to effectively maintain a secure environment and counter any threat. The U.S. conventional security capability refers to tactical U.S. elements, such as the BCT. It does not refer to higher tactical or operational headquarters or support, sustainment, or medical assets. The U.S. special operations forces capability is depicted as declining in the end of the sustainment phase; as the FSF capability improves, the need for special operations forces declines. The host-nation or regional security organization and U.S. policy determine the end state for special operations.

4-10. For figure 4-1, a major change to enemy resources may require more FSF or additional. Drastic changes to the situation may even require a new operation plan. The figure focuses on capability, not forces, so host-nation security capability may increase without increasing the number or size of host-nation forces.

4-11. Figure 4-2 illustrates how the area of operations for a modular brigade augmented for SFA may change throughout the SFA phases. The differences between the initial response phase and the transformation phase may not change on the surface, but the BCT's relationship with FSF changes drastically. For example, FSF in the north met required conditions so that one of the U.S. brigades conducting SFA was no longer required. The responsibility of the other U.S. brigade thus expanded from providing assistance to one division to two divisions. This brigade's span of control and area of operations similarly expanded. The latter stage of the transformation phase can differ greatly. Areas of operation generally increase in size as they provide SFA to more FSF.

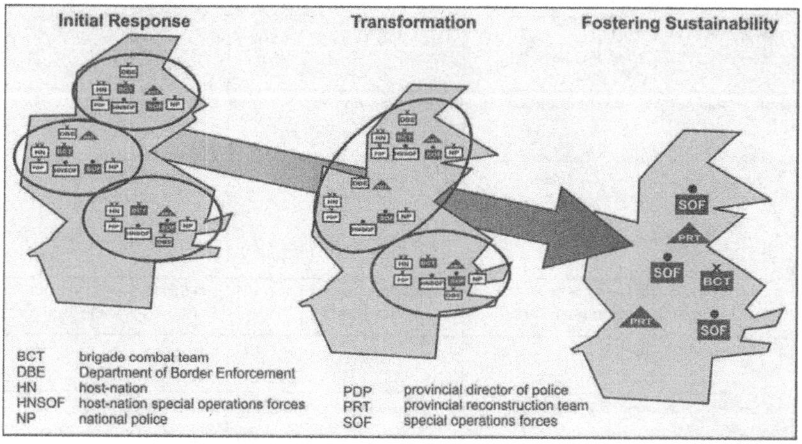

Figure 4-2. Example of changes to a brigade combat team's area of operations

AUGMENTATION BY PHASE SUMMARY

4-12. BCT augmentation is based on the operational environment and mission. As the three phases are based on the operational environment, they provide a baseline for augmentation. Potential augmentation may require military police, legal, public affairs, civil affairs, psychological operations, engineering, sociocultural experts, sustainment, and military transition team personnel. Table 4-1 illustrates a baseline for augmentation by phase for BCTs.

4-13. An embedded provincial reconstruction team is a key element of a BCT conducting security for assistance. The BCT has the lead on the following tasks: establish civil security, establish civil control (when approved by Congress), and develop and enable foreign security forces. An embedded provincial reconstruction team has the lead for the following tasks: support to economic and infrastructure development, restore essential services, and support to governance. Thus, the BCT and an embedded provincial reconstruction team are able to effectively support FSF and execute all five stability tasks.

Table 4-1. Baseline augmentation for modular brigade

CAPABILITY	INITIAL PHASE (1 HN DIV)	TRANSFORMATION PHASE (2 HN DIV)	TRANSFORMATION PHASE (3 HN DIV)
Police Provincial District	171 - MP Company 2 - MP Team (LTC/SGM) 3 - MP MAJ	171 - MP Company 4 – 2 x MP Team (LTC/SGM) 6 - MP MAJ	171 - MP Company 6 – 3 x MP Team (LTC/SGM) 9 - MP MAJ
Legal HN BDE HN DIV	2 - JAG Team (MAJ/SFC) 2 - JAG Team (CPT/SSG)	4 – 2 x JAG Team (MAJ/SFC) 4 – 2 x JAG Team (CPT/SSG)	6 – 3 x JAG Team (MAJ/SFC) 6 – 3 x JAG Team (CPT/SSG)
Public Affairs	2 - PA (Officer/NCO)	2 - PA (Officer/NCO)	2 - PA (Officer/NCO)
Civil Affairs & Psychological Operations	32 - CA Company = 5 CA Teams 70 – Tactical PSYOP Company = 9 TPTs	32 - CA Company = 5 CA Teams 70 – Tactical PSYOP Company = 9 TPTs	197 - CA Battalion (4 CA Companies) = 20 CA Teams 70 – Tactical PSYOP Company = 9 TPTs
Engineer	8 - FEST-A Team 2 - EOD (CPT/SFC)	8 - FEST-A Team 2 - EOD (CPT/SFC)	8 - FEST-A Team 2 - EOD (CPT/SFC)
Sociocultural Experts	8	8	8
Military Transition Team	56- MTT Augment	4 - MTT Augment	8 - MTT Augment
TOTAL	358	315	493
EPRT	~15-20	~15-20	~15-20
Sustainment			
Truck Squad	21 personnel		
Supply Platoon	32 personnel		
CCT	4 personnel	4 personnel	4 personnel
FMST	4 personnel	4 personnel	4 personnel

CA	civil affairs	LTC	lieutenant colonel
CPT	captain	MAJ	major
CCT	contingency contracting team	MP	military police
DIV	division	MTT	military transition team
JAG	Judge Advocate General	NCO	noncommissioned officer
EOD	explosive ordnance disposal	PA	public affairs
EPRT	embedded provincial reconstruction team	SFC	sergeant first class
FEST-A	forward engineering support team – augmenting	PSYOP	psychological operations team
FMST	financial management support team	SGM	sergeant major
HN	host-nation	SSG	staff sergeant
HTT	human terrain team	TPT	tactical psychological operations team

Initial Response Phase

4-14. The initial response phase seeks to assist FSF stabilize the operational environment in a crisis state. This phase occurs during or immediately after a conflict where the operational environment prevents civilian personnel from operating effectively. The operational environment is typified as nonpermissive.

Thus, the objective of this phase is to improve the security situation, reducing the threat to the populace and creating the conditions that allow civilian personnel to safely operate.

4-15. SFA in the initial response phase is normally required when FSF lack the capability or capacity to provide the required level of security. This phase often requires SFA efforts to help generate and train or assist new and existing FSF. This phase may require a combination of the types of SFA and considerable support, sustainment, and medical resources. BCT activities during the initial response may have to be conducted with multinational major combat operations, to include providing a safe, secure environment for the local populace.

4-16. The BCT's SFA efforts during this phase focus on improving the foreign security force's capability and capacity so all security forces—U.S., other, and foreign security forces—provide a secure environment and reduce the threat. Figure 4-3 illustrates the BCT's span of control. The figure also depicts three BCTs' operational areas in the initial response phase. As security conditions improve, transition to the transformation phase begins.

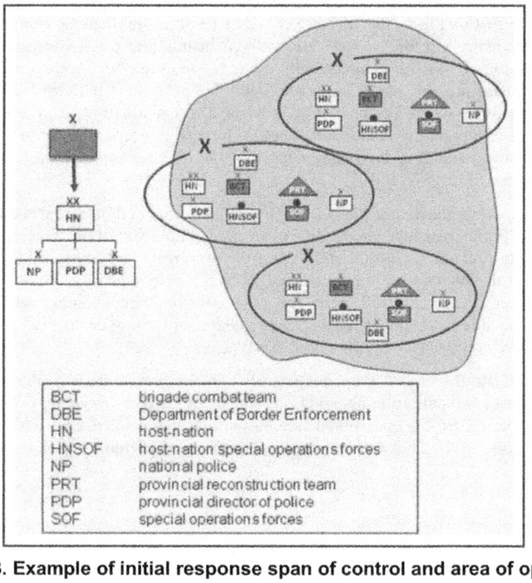

Figure 4-3. Example of initial response span of control and area of operations

4-17. The BCT requires augmentation for the initial response phase. These embedded units provide key enabling capabilities for the brigade, supported FSF, and other supported actors. The embedded provincial reconstruction team is one of the most important of the brigade's augmentation, as it provides interim civilian expertise in developing local institutions to take the lead in national governance, the provision of basic services, fostering economic development, and enforcement of the rule of law. Column two of table 4-1 summarizes augmentation during the initial phase. Although each brigade's mission and conditions in its area of operations are unique, the baseline augmentation for the modular brigade operating in this phase is as follows:

- Embedded provincial reconstruction team.
- Forward engineering support team-augmenting.
- Explosive ordnance disposal team.
- Sociocultural experts.

- Military police company (provincial police transition team).
- Military police team (district police transition team).
- Three military police majors (district police transition team).
- Two judge advocate general teams.
- Civil affairs company.
- Tactical psychological operations company.
- Public affairs team.
- Contingency contracting team.
- Financial management support team.
- Military transition team (specific augmentation personnel include two lieutenant colonels, thirty-one majors, and twenty-three master sergeants).

TRANSFORMATION PHASE

4-18. In the transformation phase, the BCT seeks to assist FSF to stabilize the operational environment in a crisis or vulnerable state. The operational environment in this phase is more permissive than the initial response phase; however, military forces will often be required to provide security to some actors. Activities in this phase normally include a broad range of post-conflict reconstruction, stabilization, and capacity-building efforts, which the embedded provincial reconstruction team is essential for long-term success. Objectives in this phase include continuing efforts to improve the security situation, reducing the threat to the populace, building host-nation capacity across the stability sectors, and facilitating the comprehensive approach to assist FSF.

4-19. The transformation phase represents a broad range of SFA activities that the BCT could perform to support FSF. The initial response phase differs from the transformation phase in the foreign security force's capability to provide for a safe and secure environment. More specifically, FSF have a level of proficiency so they no longer need a permanent U.S. and foreign security forces relationship for tactical operations. However, they may still need full-time advisors and support, sustainment, and medical assistance. The embedded provincial reconstruction team will continue to play a vital role in assisting governance and development efforts throughout this phase.

4-20. Figure 4-4 illustrates the BCT conducting SFA in the transformation phase. In this example, the BCT's span of control expands from assisting one foreign division to assisting two foreign divisions. The figure also depicts how a BCT's operational area has similarly expanded. SFA activities in this phase seek to establish conditions so the host nation's security sector can provide a secure environment with its own security forces.

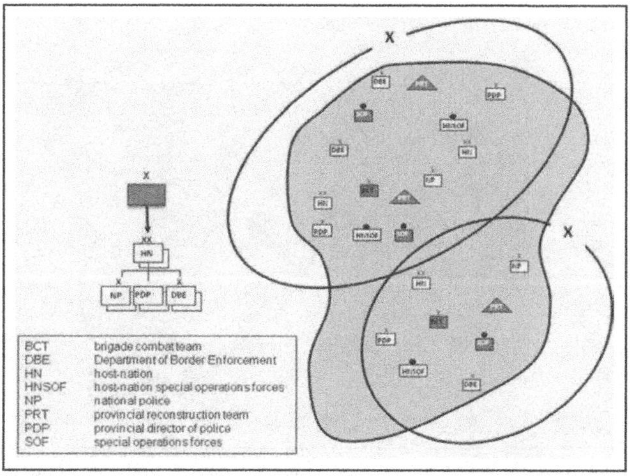

Figure 4-4. Example of transformation phase span of control and area of operations

4-21. The BCT requires augmentation for the transformation phase, like the initial response phase. It often requires similar augmentation due to its greater span of control and area or operations. Column three of table 4-1 summarizes augmentation during the transformation phase with two foreign divisions. The baseline augmentation for the brigade in this phase is as follows:

- Embedded provincial reconstruction team.
- Forward engineering support team-augmenting.
- Explosive ordnance disposal team.
- Sociocultural experts.
- Military police company (provincial police transition team).
- Two military police teams (district police transition team).
- Six military police majors (district police transition team).
- Four judge advocate general teams.
- Civil affairs company.
- Tactical psychological operations company.
- Public affairs team.
- Contingency contracting team.
- Financial management support team.
- Military transition team (specific augmentation personnel include two lieutenant colonels and two master sergeants).

4-22. As conditions of the operational environment improve, the BCT may require a transition to assist three foreign divisions, rather than two divisions as depicted in figure 4-4. This represents a significant expansion of the span of control and area of operations, with a corresponding expansion of required augmentation. Figure 4-5 depicts the BCT supporting three foreign divisions.

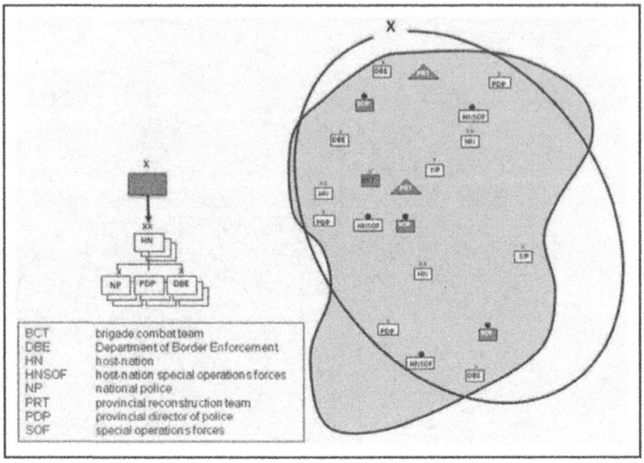

Figure 4-5. Example of a three-division span of control and area of operations

4-23. As in other phases, the BCT requires augmentation to support three divisions. Column four of table 4-1 summarizes augmentation during the transformation phase with three foreign divisions. The baseline augmentation for the brigade in this phase is as follows:

- Embedded provincial reconstruction team.
- Forward engineering support team-augmenting.
- Explosive ordnance disposal team.
- Sociocultural experts.
- Military police company (provincial police transition team).
- Three military police teams (district police transition team).
- Nine military police majors (district police transition team).
- Six judge advocate general teams.
- Civil affairs company.
- Tactical psychological operations company.
- Public affairs team.
- Contingency contracting team.
- Financial management support team.
- Military transition team (specific augmentation personnel include four lieutenant colonels and four master sergeants).

FOSTERING SUSTAINABILITY PHASE

4-24. In this phase, the focus of SFA continues to shift toward assisting institutions required to sustain FSF. This phase encompasses long-term efforts to assist FSF. In this phase, FSF are capable and viable. They conduct independent operations and can provide a safe, secure internal environment. While the brigade may be initially required during this phase, it is normally withdrawn during this phase. The determination to withdraw the brigade is based on the policy and conditions of the operational environment. Special operations forces, provincial reconstruction teams, and other forces may remain to support a theater security cooperation plan. Thus, remaining SFA missions are normally special operations forces missions, not a BCT mission (special operations forces will normally be involved in all phases of SFA).

TRANSITIONS

4-25. SFA can require a vast array of transitions. These transitions are conducted according to the tactical, operational, and strategic conditions identified during planning (which must support the internal defense and development strategy and using an overall comprehensive approach). Transitions can occur simultaneously or sequentially in different levels or war and in separate echelons. This includes potentially having tactical transitions for different units within one brigade's area of operations.

4-26. There are several major transitions. First, the supported unit may be the U.S. BCT in the beginning of an initial response phase but the foreign security forces will transition to be the supported unit. At this point in the transformation phase, the area in which the BCT conducts SFA will expand. This expansion can occur multiple times during the transformation phase, which is based on conditions, especially the capability and capacity of FSF.

4-27. Operations never proceed exactly as planned. To facilitate flexibility, commanders incorporate branches and sequels into the overall plan. Visualizing and planning branches and sequels are important because they involve transitions. Unless planned, prepared for, and executed efficiently, transitions can reduce the tempo of the operation, slow its momentum, and surrender the initiative to the adversary.

SUBORDINATE UNITS

4-28. The modular brigade augmented for SFA has subordinate units whose sole focus is working with FSF. These advisor teams may be formed from brigade organic resources, external augmentation, or a combination. These teams optimally are embedded with the counterpart unit, or they may reside on a U.S. camp and commute to FSF they support. The method depends on U.S. policy, direction from higher headquarters, the conditions of the operational environment, and capacity of the foreign security force's camps to accommodate the U.S. forces.

4-29. The military transition team (MTT) is a key subordinate unit for the brigade conducting SFA. The MTT mission is to assist FSF units. Figure 4-6 illustrates the relationship between a brigade conducting SFA and a MTT assisting a foreign brigade. In this example, the brigade has a battalion and a division MTT as subordinate units, although it would have other subordinate units. The brigade MTT is subordinate to the battalion and has operational control of one of the parent U.S. battalion's company teams.

4-30. When the brigade provides organic forces to form the basis for the brigade MTT, the company team is the foundation of a brigade MTT. The company team will be augmented with additional personnel and assets. The U.S. brigade must facilitate the MTT operations, such as force protection, transportation, communications, and reachback.

4-31. As companies do not have organic intelligence analysts, personnel must be used to form company intelligence support teams (COISTs). A COIST, an ad hoc organization, assists the MTT and its FSF. Ideally, the COIST should consist of six individuals, which will allow for 24 hour operations and the depth in the team to both integrate with the staff and to successfully support the company's mission. The team should consist of an officer, a noncommissioned officer, and four Soldiers, although the cell can be minimally manned with four personnel. The company intelligence cell works most effectively if its work complements the battalion S-2's work.

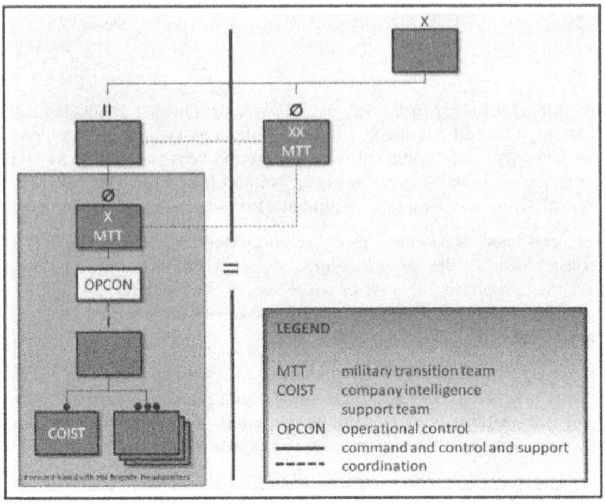

Figure 4-6. Military transition teams

SAMPLE TROOP-TO-TASK MODEL

4-32. The BCT may support assigned to multiple missions, including potentially supporting multiple foreign units in their operational area. Additionally, these FSF organizations may each report through different host-nation government channels and even to different ministries. To synchronize efforts in this case, U.S. forces must achieve unity of command and effort. Similarly, each of the FSF organizational commanders should synchronize their efforts with the host-nation government representatives, as appropriate.

4-33. Figure 4-7 depicts a sample troop-to-task model for a U.S. brigade. It depicts one company team acting as a quick reaction force; thus, the battalion retains the capability to conduct independent full spectrum operations. The figure depicts how U.S. company teams may form the basis of a brigade MTT or a brigade border transition team (and subordinate point of entry transition teams). The figure also depicts how a company team may provide police support. This support is in the form of two platoons supporting police transition teams and a third platoon in a joint security station providing support to a police transition team.

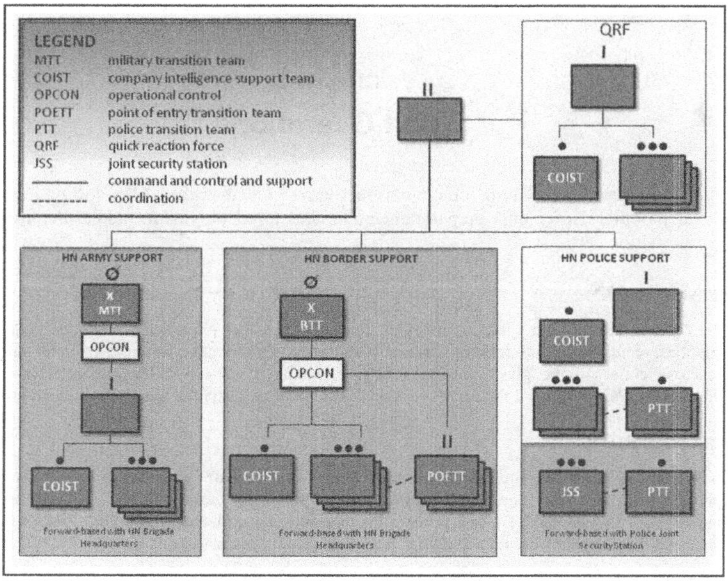

Figure 4-7. Battalions conducting security force assistance

4-34. Subordinate units of BCTs conducting SFA are best located inside a FSF base—not an American base. Co-location facilitates integration with the FSF. It allows the two forces to form mutual understanding and trust. Co-location and the close cooperation it facilitates often improves the population's perception of the legitimacy U.S. and foreign security forces, which can be an essential condition of the overall mission's end state.

4-35. If force protection requires, a U.S. area may be established in the FSF base, although this is not optimal. Key considerations for co-location may include the threat, FSF acceptance, physical space inside the FSF base, sustainment capabilities, medical facilities, and availability of quick reaction forces.

4-36. When U.S. forces are operating out of smaller outposts in an urban environment, the local populace sees the integration and presence of the U.S. and foreign security forces working together. This integration not only enhances overall operational effectiveness and trust, living and working together builds legitimacy of the two forces as well as FSF; it reinforces trust between the FSF and the people they are tasked to protect.

Chapter 5
Unit Operations

Unit operations require continuous activities. Commanders plan for premission, deployment, in-country preparation, employment, redeployment, and postdeployment activities.

PREMISSION

5-1. After receiving a mission, the unit continues detailed preparation activities. The unit obtains supplies, equipment, and training materials and prepares for deployment. Personnel prepare for and rehearse classes to be given in-country. They hone their military skills and conduct extensive area orientation. Paragraphs 5-2 through 5-10 discuss staff actions particular to security force assistance (SFA).

OPERATION PLAN

5-2. The S-3 disseminates the operation plan to unit staff and subordinates. The S-3 ensures predeployment training for Soldiers, to include preparation for training foreign security forces (FSF) and rehearsals for movement. The S-3 reviews the program of instruction for training FSF, to include getting approval from the commander and higher headquarters, if necessary.

CIVIL-MILITARY OPERATIONS

5-3. The civil-military operations section ensures the operation plan minimizes how operations affect the civilian population and addresses ways to mitigate the civilian impact on military operations. Civil-military operations also ensure the operation plan discusses coordination with any augmenting actors. These actors might include civil affairs battalions or brigades, provincial reconstruction teams, or United States Agency for International Development project officers in the operational area. During predeployment training, Soldiers receive training, materials, and briefings on the operational area. This training can cover the history, culture, religion, language, tribal affiliations, local politics, and cultural sensitivities as well as any significant nongovernmental organizations operating in the operational area.

TRAINING

5-4. As discussed in chapter 2, advisors should focus their premission training on the specific requirements of developing FSF. The training emphasizes the host-nation culture and language and provides cultural tips for developing a good rapport with foreign personnel. All premission training for SFA follows the training principles described in FM 7-0.

5-5. Based on the higher commander's training guidance, unit commanders assign missions and approve the draft mission-essential task list that supports SFA. The staff plans, conducts, and evaluates training to support this guidance and the approved mission-essential task list for SFA missions. Commanders prioritize tasks that need training. Since there is never enough time to train in every area, commanders focus on tasks essential for mission accomplishment.

5-6. Once commanders select tasks for training, the staff builds the training schedule and plans on these tasks. The staff provides the training requirements to the commander. After approving the list of tasks to be trained, the commander includes the tasks in the unit training schedule. The staff then coordinates the support and resource requirements with the S-3 and S-4. Finally, the commander ensures standards are enforced during training.

Evaluation

5-7. Evaluations can be either internal or external. Internal evaluations occur at all levels, and they must be inherent in all training. External evaluations are usually more formal and conducted by a headquarters two levels above the unit being evaluated. This subject must be carefully planned and discussed with FSF leaders to account for cultural sensitivities and current capabilities.

5-8. A critical weakness in training is the failure to evaluate each task every time it is executed. Every training exercise provides potential for evaluation feedback. Every evaluation is also a training session. Leaders and trainers must continually evaluate to optimize training. Evaluation must occur as training takes place. Emphasis is on direct, on-the-spot evaluations. However, leaders allow Soldiers to complete the task first. Leaders plan after action reviews at frequent, logical intervals during exercises. This technique allows the correction of shortcomings while they are still fresh in everyone's mind. The after action review eliminates reinforcing bad habits.

Specified Training

5-9. Augmentation elements require area orientation, refresher combat training, field training exercises, and the like. Unit training objectives are for developing capabilities to conduct internal defense and development (IDAD) activities for tactical operations, intelligence operations, psychological operations, populace and resources control operations, and civil affairs and advisory assistance operations in the host-nation language. Units identified for SFA begin intensified training immediately upon deployment notification.

5-10. After deployment to the host nation and before commitment to operations, the unit may receive in-country training at host-nation training centers or at designated training locations. This training helps personnel become psychologically and physically acclimated to the host nation. This training also allows commanders and staff some time to coordinate and plan within their own command and with civilian and military joint and multinational organizations. After commitment, training continues and is stressed between operations, using needed improvements identified in operations as the basis for training.

DEPLOYMENT

5-11. SFA is often conducted in semipermissive environments in which U.S. forces are guests of the host-nation or partner organization. Units move into the operational area by following their deployment standing operating procedures and the operation plan. Paragraphs 5-12 through 5-25 highlight organizations and procedures critical to SFA deployment. This information, however, does not obviate the need for extensive standing operating procedures and an operation plan.

PREDEPLOYMENT SITE SURVEY

5-12. The predeployment site survey unit aims to report accurately to its parent unit the existing FSF mission variables—mission, enemy, terrain and weather, troops and support available, time available, civil considerations (METT-TC). See table 5-1. It also establishes in-country command and control, support, protection, and sustainment relationships for the follow-on unit's mission execution. The site survey coordinates the in-country reception of SFA forces. This coordination wards off most friction during the handoff. This coordination requires all units—outgoing and incoming—to apply a comprehensive approach at all levels of command. Appendix D of FM 3-05.202 discusses site survey procedures in detail.

Table 5-1. Advisor team predeployment site survey checklist

Tasks	*Characteristics*
Area of operations overview	• Map familiarization. • In sector orientation. • Areas of emphasis.
Graphic control measures	• Unit boundaries. • Routes. • Other United States and multinational support locations. • Maintenance support location. • Medical support locations. • Other points of interest.
United States and multinational unit interviews by specialty staff	• Responsibilities. • Battle rhythm. • Required reports. • Standing operating procedures, continuity book. • Stay-behind equipment.
Briefings on advised unit	• Foreign security forces task organization. • Foreign security forces key leadership. • Intelligence assessment. • Unit operational assessment.
Advised unit introductions	• Among commanders. • Among staffs.

ADVANCE PARTY DEPLOYMENT

5-13. An advance party deploys to the host-nation or partner organization area of operations after the unit completes its mission planning and the proper commander representative declares it deployable. The advance party conducts predeployment activities and coordinates necessary travel documentation and country clearances. The advance party reviews the rules of engagement and any status-of-forces agreements before deploying. If applicable, the proper staff provides an updated threat briefing to the unit. The S-4 section and support company help with out-loading in the unit area. The unit places all supplies and materials on pallets beforehand and prepares the load manifest.

5-14. Upon arrival, the advance party processes through customs and immediately notifies the higher headquarters of its arrival and status. Often the higher headquarters provides point of contacts to the advance party. These contacts help the advance party obtain the proper identification, documentation, and weapons permits.

5-15. The advance party establishes contact with all U.S. and foreign security forces tactical unit commanders and conducts final premission planning activities. The advance party procures secure working, storage, and living areas for the unit and reconnoiters and prepares the training sites. The advance party coordinates for special support equipment to unload heavy supplies and transport the supplies and personnel to the training site.

MAIN BODY DEPLOYMENT

5-16. The unit deploys after completing its planning and the appropriate commander declares the unit ready. The unit follows the same procedures performed by the advance party. Members of the advance party meet the main body on its arrival and brief the main body on any required changes to the operation order.

5-17. After arriving in-country, the unit processes through customs and immediately notifies the higher headquarters of its arrival and status. The unit obtains the proper identification, documentation and weapons permits from U.S. or other sources, if applicable. Often the security assistance organization provides to the unit the necessary point of contacts that expedite this process.

5-18. The unit off-loads equipment and personnel onto vehicles for movement to the training site, or it stores the equipment in a secure area until the processing is completed. FSF secure the equipment but the unit members remember the commander is ultimately accountable for the advisor team's physical and personal security. Supply personnel stay with the equipment and help to guard it.

IN-COUNTRY PREPARATION

5-19. Upon arrival, the commander and S-3 brief the higher headquarters on the planned execution of the mission and reconfirm the required command relationship. Local conditions may require the unit to confirm or establish its in-country and external command and control, support, and sustainment relationships from outside its operational area upon arrival. The unit establishes direct working relationships with its next higher in- or out-of-country supporting element to—

- Determine the limits of the available support and expected reaction time between the initiation of the support request and its fulfillment.
- Confirm or establish communications procedures between the supporting element and the unit, to include alternative and emergency procedures for command and control, all available support, and medical evacuation.

5-20. The unit establishes procedures to promote interagency cooperation and synchronization. The unit—

- Identifies the location of the concerned host-nation, U.S., or other agencies.
- Contacts the concerned agency to establish initial coordination.
- Exchanges information or intelligence.
- Confirms or establishes other coordination protocols as necessary.
- Incorporates the newly established or changed procedures into the plans for mission execution.

5-21. The unit immediately establishes operations security procedures to support its mission execution and identifies rally points incorporated into its defensive, evasion, and personnel recovery plans.

5-22. After receiving a detailed briefing and further guidance from the advance party, unit personnel continue to develop effective rapport with the FSF commander and counterparts. They also assess their working, storage, and living areas for security and verify the location of the training site, communications center, dispensary area, and FSF troop area. With the FSF commander, the unit commander—

- Establishes rapport.
- Conducts introductions in a businesslike, congenial manner.
- Briefs on the unit's mission, its capabilities, and the restrictions and limits imposed on the detachment by the higher U.S. commander.
- Ensures all unit personnel fully support FSF and firmly believes a joint U.S.-FSF effort will succeed. Requests counterpart linkup be made under the mutual supervision of the FSF commander and the unit commander.
- Ensures all current unit plans are tentative and that assistance is needed to finalize them.
- Deduces or solicits the actual estimate of unit capabilities and perceived advisory assistance and material requirements.
- Recommends the most desirable courses of action (COAs) while emphasizing how they satisfy present conditions, achieve the desired training, and meet advisory assistance goals.
- Informs the higher in-country U.S. commander of any significant changes in the unit's plan to assist FSF.

5-23. The S-2 supervises the dissemination of intelligence and other operationally pertinent information within the unit and, as applicable, to higher, lower, or adjacent units or agencies. The S-2 also—

- Monitors the implementation of the intelligence collection plans to include updating the commander's priority intelligence requirements (PIRs), conducting area assessment, and coordinating for additional intelligence support.
- Establishes liaison with FSF intelligence and security agencies (within the guidelines provided by applicable higher authority). Assesses the intelligence threat and resulting security requirements, including coordination with the S-3 on specific security and operations security measures.

5-24. Through the S-2, the commander's PIRs are based on the latest information available and requirements for additional PIRs that arise from modified estimates and plans. The S-2 also—

- Analyzes the foreign unit's status to finalize unit plans for advisory assistance. These plans can include task organization of unit with counterparts, staff functions for planning SFA, and advisory assistance for executing SFA.
- Explains analysis to counterparts and encourages them to help with—and participate in—analyzing, preparing, and briefing the analysis to the foreign unit commander.
- Prepares and briefs the plans for training and advisory assistance.
- Helps the foreign unit inspect the available facilities to identify deficiencies. If the unit finds deficiencies, the S-2 prepares estimates of COAs for the FSF commander to correct them.
- Supervises the preparation of the facilities with their counterparts and informs unit and FSF commanders on the status of the facilities.

5-25. The unit ensures its security is based on the present or anticipated threat. Some recommended actions the unit may take include—

- Hardening its positions based on available means and requirements to maintain low visibility.
- Maintaining unit internal guard system with at least one Soldier who is awake and knows the locations of all other unit personnel. The guard reacts to an emergency by following an internal alert plan and starting defensive actions.
- Maintaining communications with all subordinate unit personnel deployed outside the immediate area controlled by the main body.
- Establishing plans for immediate defensive actions in the event of an attack or a loss of rapport with hostile reaction.
- Discussing visible security measures with foreign counterparts to ensure understanding and to maintain effective rapport. Unit personnel do not divulge sensitive information for the sake of possible rapport benefits.
- Encouraging the foreign unit, through counterparts, to adopt additional security measures identified when analyzing the foreign unit's status and inspecting its facilities.
- Coordinating defensive measures with the foreign unit to develop a mutual defensive plan. Unit personnel obtain from the unit's present reaction and defensive plans for attack. They encourage the foreign unit to conduct mutual full-force rehearsals of defensive plans; if unsuccessful, the unit conducts internal rehearsals of the plans.

EMPLOYMENT

5-26. Once the unit has deployed and arrived in-country, it begins employment. Employment is the conduct of SFA to support the FSF commander. Employment occurs within the advisor team itself. It includes foreign counterparts in the mission planning (preparing the FSF for the mission itself) to increase the capability and capacity of their planning processes, as well as increase the probability of success.

PERSONNEL

5-27. During employment, individual unit personnel perform functions described in paragraphs 5-28 through 5-38.

Commander

5-28. Before the mission, the U.S. commander advises and assists the FSF commander. The latter issues planning guidance for planning the execution of the mission and clarifies commander's intent. The U.S. commander advises and assists the FSF unit commander in the command and control process of tactical operations. By accompanying the FSF unit commander when the mission is received from higher headquarters, the U.S. commander assists any subsequent missions. The U.S. commander monitors how FSF understand the commander's intent and all specified or implied tasks.

5-29. During execution of the mission, the U.S. commander helps the foreign unit commander provide command and control during operations. After monitoring the tactical situation, the U.S. commander recommends changes to the COAs to exploit the situation. After monitoring the flow of information, the U.S. commander recommends improvements to use intelligence collection assets and to keep subordinates reporting required information.

Staff

5-30. Before the mission, the advisor staff advises and assists the foreign counterparts in preparing estimated COAs for essential tasks. The staff helps write tentative plans based on the planning guidance and unit's standing operating procedures. These plans include primary, alternate, contingency, and emergency plans.

5-31. During execution, the staff helps foreign counterparts coordinate the execution of the tasks. They disseminate portions of their plans to other personnel, senior and subordinate staff sections, and supporting elements. They help notify higher, lower, or adjacent staff sections of modified estimates and plans. Finally, the staff—with the S-2 and the S-2 counterpart—helps update the commander's critical information requirements with the latest information and requirements that arise from operations.

Executive Officer

5-32. The executive officer performs the organizational analysis of the unit's coordinating staff sections to ensure efficiency during the planning process according to initial planning guidance. With the foreign counterpart, this officer directs foreign staff sections as they develop estimates and plans. The executive officer monitors the liaison and coordination with FSF higher headquarters, recommending changes to improve efficiency.

Adjutant

5-33. The adjutant provides advice, assists, and makes recommendations to the foreign counterpart for conducting human resources. This includes monitoring the maintenance of foreign unit strength, accountability of casualties, unit morale, and postal activities. This may also include concerns with the foreign pay system, leave procedures, and casualty pay procedures.

Intelligence Officer

5-34. The intelligence officer advises and assists the monitoring of FSF operations security to protect classified and sensitive material and operations and recommends improvements. By helping the foreign counterpart update the situation map, the intelligence officer helps to keep the map current. The intelligence officer recommends improvements to the standing operating procedures of the tactical operations center communications so the intelligence section receives situation reports. This officer helps the counterpart monitor the collection, evaluation, interpretation, and dissemination of information. The intelligence officer assists in the examination of captured insurgent documents and material. This officer helps gather and disseminate intelligence reports from available sources to ensure the exploitation of all assets. The intelligence officer helps the counterpart to brief and debrief patrols operating as a part of reconnaissance and surveillance activities. Finally, the intelligence officer assists, with the advisor operations officer, developing reconnaissance and surveillance plans.

Operations Officer

5-35. The operations officer helps the foreign counterpart to prepare tactical plans using estimates, predictions, and information. This officer monitors command and communications nets, assists in preparing all orders and plans, and helps to supervise the training and preparation for operations. Finally, the operations officer monitors the planning process and makes recommendations for consistency with IDAD goals.

Supply Officer

5-36. The supply officer advises and assists the foreign counterpart in maintaining equipment readiness; monitoring the support provided to the foreign unit, its subunits, and attachments; and recommending improvements. Finally, this officer helps to supervise the use of transportation assets.

Civil-Military Team

5-37. Upon deployment, civil-military teams advise the commander and staff on civil-military considerations and coordinate efforts of any civil affairs units supporting the unit. The civil-military team mentors the counterpart team on the supported foreign element staff on civil-military operations, the legal and moral obligations of military commanders to civilian populations under their control, and the importance of respecting human rights. The civil-military team may introduce the counterpart to relevant nongovernmental organizations, United States Agency for International Development project officers, and provincial reconstruction team staff.

Company-Level Advisors

5-38. Advisors sent to elements below battalion level help foreign counterparts to analyze the mission and commander's intent from higher headquarters. They assist FSF leaders restate the mission, conduct an initial risk assessment, identify a tentative decisive point, and define their own intent. They also assist their foreign counterparts to analyze the mission variables. From these variables, advisors help their foreign counterparts to develop a COA that meets the higher headquarters concept of operations and commander's intent. Finally, they advise and assist in the conduct of operations and the flow of information to the FSF higher commander.

TRAINING

5-39. Training FSF during SFA parallels the training of U.S. forces. Training follows the principles of training described in FM 7-0. This training tailors the objectives to account for cultural differences and focus on the FSF mission and problem set.

Training Assessment

5-40. The commander of the advising unit begins with a training assessment of training plans designed during predeployment and passed on from the preceding U.S. units conducted in coordination with the FSF commander. This assessment is important to both evaluate the foreign security force but also as an exercise to establish and enhance the working relationship between U.S. and foreign security forces. The training assessment should cover all aspects of leadership, training, sustainment, and professionalization. To support an assessment, the advising unit should analyze the following specific foreign unit considerations:

- The unit's mission and mission-essential task list and capability to execute them.
- Staff capabilities.
- Personnel and equipment authorization.
- Physical condition.

- Any past or present foreign influence on training and combat operations.
- Operational deficiencies identified during recent operations or exercises with U.S. personnel.
- Sustainment capabilities, to include sustainment training programs.
- Internal training programs and personnel.
- Training facilities.

5-41. The level of professionalism of FSF, both units and individuals, should also be assessed. Adhering to established rules of engagement, ethics that meet the established laws and regulations of the commanding authority, laws for land warfare, and human rights, promotions based on merit and the security forces' support of civilian leaders and political goals all fall within this assessment.

5-42. The advisor unit, working with FSF leaders, also evaluates current members of the security forces for past military skills and positions. Often military reorganizations arbitrarily shift personnel to fill vacancies outside their knowledge and experience.

Training Plan

5-43. After completing the training assessment, advisors analyze the prepared training plan and determine if changes are necessary. The training plan stresses the deficiencies identified in the training assessment. It also identifies those in the host nation able to help train FSF and to strengthen the legitimacy of the process. Finally, the training plan should consider the unit's eventual self-sustainment. As the foreign security force gains sufficient capacity and capabilities to perform independently, advisors transition from a leading role to a mentoring role.

5-44. The plan for training FSF uses a comprehensive approach, supports the IDAD strategy, and has many echelons. Working with other echelons and actors within the area of operations can provide support and expertise that enhance the training and operations process as well as limit redundancy.

Program of Instruction

5-45. In coordination with the FSF commander and training staff, the advising unit develops a program of instruction. This program incorporates all training objectives that satisfy the training requirements identified during the assessment. The training program must support these requirements. The FSF commander approves the program of instruction prior to execution. When executing the program of instruction, advisors adhere to a training schedule consistent with changes in the mission variables. Advisors ensure through their counterparts and the FSF commander that all personnel receive training. Foreign counterpart trainers rehearse all classes approved on the program of instruction.

Presentation of Instruction

5-46. To present the training material properly, trainers follow the lesson outlines approved in the program of instruction. All training clearly states the task, conditions, and standards desired during each lesson, ensuring the foreign students understand them. Trainers state all warning and safety instructions in the host-nation language. The training to reinforce the concepts includes demonstrations of the execution of each task, stressing the execution as a step-by-step process. Trainers monitor students' progress during instruction and practical exercises, correcting mistakes as they are made.

Training Methods

5-47. An effective method of training used by U.S. forces is the crawl-walk-run method of teaching individual tasks, battle drills, collective tasks, and field exercises. (See FM 7-0.) This method is employed to develop well-trained leaders and units. During all phases, the training must include the mission of the unit in the context of the higher unit's mission to assist with the practical application of the training. Identifying the higher commander's mission and intent, as well as the tasks and purposes of other units in the area, would also add context to the training. This method can also be expanded to include the role of other actors.

5-48. At a minimum, units continue individual training to improve and sustain individual task proficiency while training on collective tasks. Collective training requires interaction among individuals or organizations to perform tasks, actions, and activities that contribute to achieving mission-essential task proficiency. Collective training includes performing collective, individual, and leader tasks associated with each training objective, action, or activity.

Collective Training

5-49. Collective training starts at the squad level. Squad battle drills provide key building blocks that support operations. Trainers can link battle drills and collective tasks through a logical, tactical scenario in a situational training exercise. Although this exercise is mission-oriented, it results in more than mission proficiency. Battle drills and collective tasks support situational training exercises, while these exercises support operations. Advisors must understand the operational environment when training FSF; training incorporates how internal and external threats and civilians affect the environment. Flexibility in using U.S. doctrine in training enhances efforts to make training realistic. Trainers modify U.S. doctrine to fit the FSF level of expertise, their command and control, the tactical situation, and sustainment base. Often the structure and capabilities of FSF differ from that required by U.S. doctrine. When FSF counter an insurgency, these exercises emphasize interplay among psychological and tactical, populace and resources control, intelligence, and civil affairs operations. Similarly, scenarios for each tactical exercise must include interplay among the various possible aspects of IDAD.

Individual Training

5-50. Training individuals of FSF emphasizes physical and mental conditioning, tactical training, basic rifle marksmanship, first aid, combatives, and the operational environment. Individual training also includes general tactics and techniques of security operations and the motivation, operations, and objectives of internal and external threats. Tough and realistic training conditions troops to mentally and physically withstand the strain of continuous offensive operations.

5-51. Foreign personnel cross-train on all types of weapons, communications and other equipment, and skills particular to their unit. Personnel losses must never cause weapons, communications equipment, or essential skills to be lost due to a lack of fully trained replacement personnel.

Small-Unit Leader Training

5-52. SFA frequently entail rapidly changing circumstances; thus, FSF small-unit leaders must be able to plan and execute operations with little guidance. Therefore, trainers stress small-unit leadership training which can occur concurrently with individual training. Tools the trainer uses to train leaders are manuals, previously established training, tactical exercises without troops, battle simulations, and unit missions. This training develops aggressiveness, tactical proficiency, and initiative. Small-unit leader training should include combined arms technical training such as procedures for forward observer and close air support. Leadership training must also include land navigation in difficult terrain and under conditions of limited visibility. Mission readiness and the health and welfare of subordinates are also continuous parts of training.

After Action Review

5-53. In any training, the after action review provides the critical link between training and evaluation. It is a professional discussion that includes the training participants and focuses directly on the training goals. An after action review occurs after all collective foreign unit training. Effective after action reviews review training goals with the concerned FSF unit commander. Advisor personnel ask leading questions, surface important tactical lessons, explore alternative COAs, assist the retention teaching points, and keep the after action review positive.

5-54. The advisor encourages the FSF unit commander to review the training with the entire unit or key subordinate leaders. At this review, the FSF commander stresses how to strengthen the chain of command. During the after action review, advisors avoid criticizing or embarrassing the FSF unit commander or subordinates.

5-55. If possible, after action reviews occur during the field portion of the training when the unit assembles at logical stopping points. These reviews provide feedback from instructors and leaders that increases and reinforces learning. This feedback provides a richer database for key points. During the after action review, the senior evaluator draws information from unit leaders to form possible alternative COAs.

Assessment

5-56. In all operations, including SFA, success is generally measured for three periods: short-, mid-, and long-term. Success in SFA is normally defined within the context of these three periods. In the short-term period, FSF make steady progress in fighting threats, meeting political milestones, building democratic institutions, and standing up security forces. In the mid-term period, FSF lead fighting threats and provide security, have a functioning government, and work towards achieving economic potential. In the long-term period, FSF are peaceful, united, stable, and secure; integrated into the international community; and a full partner in international security concerns.

Measures of Effectiveness

5-57. A *measure of effectiveness* is a criterion used to assess changes in system behavior, capability, or operational environment that is tied to measuring the attainment of an end state, achievement of an objective, or creation of an effect (JP 3-0). A measure of effectiveness provides a benchmark against which the commander assesses progress toward accomplishing the mission.

5-58. The following are tools to assess the effects of operations in meeting expectations defined for the short, mid, and long term. Although all SFA is not focused on a host nation, these tools are based on the four IDAD pillars:

- **Balanced development** attempts to achieve national goals through political, social, and economic programs.
- **Security** includes all activities implemented to protect the populace from the threat and to provide a safe environment for national development.
- **Neutralization** is a political concept that encompasses the physical and psychological separation of the threatening elements from the population.
- **Mobilization** provides organized manpower and resources and includes all activities to motivate and organize popular support of the government.

Measures of Performance

5-59. A *measure of performance* is a criterion used to assess friendly actions that is tied to measuring task accomplishment (JP 3-0). This quantitative analysis determines whether the task or action was performed to standard. The standard may be derived from a procedure or may be time-sensitive. Examples of measures of performance are listed in the appropriate Army tasks in FM 7-15.

5-60. The advisor helps the foreign counterpart to monitor the current situation for unanticipated successes, failures, or enemy actions. As the commanders assess the progress of foreign security operations, they look for opportunities, threats, and acceptable progress. Throughout the operation, the commanders visualize, describe, and direct changes to the operation, feeding the assessments of progress or regression back into the planning process.

REDEPLOYMENT

5-61. Redeployment is the return of forces and materiel to the home or mobilization station. Before redeployment, a unit conducting SFA inventories all supplies and equipment (jointly, if possible) to be passed on to the foreign unit or the relieving U.S. or multinational unit. These actions ensure all items are accounted for and the custody chain for property and equipment is unbroken. The outgoing unit also—

- Prepares a final evaluation report showing its estimate of the foreign unit's capabilities and giving an opinion of its future employment.
- Discusses the foreign unit's performance with its commander. The unit submits a copy of the foreign unit's final evaluation to the next higher FSF commander.

- Passes custody of training schedules, lesson plans, foreign operational records, and the foreign unit's final evaluation to the foreign unit and relieving U.S. or coalition unit.
- Presents debriefings, after action reviews, and the foreign unit's final evaluation to the higher in-country U.S. commander.

5-62. The outgoing unit commander supervises the redeployment. This commander approves prepared redeployment plans or, as necessary, directs changes to the redeployment plans. The commander and operations officer assess and select alternatives or changes to the present redeployment COAs based on the intelligence officer's intelligence estimate. They develop and disseminate a fragmentary order for selected alternatives or changes.

5-63. Unit personnel prepare all accompanying supplies and equipment for shipment. They brief ground support personnel on equipment handling procedures, routes, convoy procedures, and actions to take if a terrorist or insurgent incident occurs. They load personnel and equipment for movement to the departure airfield or airport following the load plan in the unit's operation order. They maintain accountability for all their personnel, supplies, and equipment.

Mission Handoff Procedures

5-64. During long-term SFA, conditions determine the rotation of in-theater units. Time is not the only governing factor. Changes in the operational environment may require reshaping force packages as situations change. In addition, internal administrative concerns might prompt or support a commander's decision to rotate units. Regardless, mission handoff is necessary and defined as the process of passing an ongoing mission from one unit to another with no discernible loss of continuity.

5-65. The overall authority for the handoff and the subsequent transfer of authority lies with the commander ordering the change. The authority for determining the handoff process lies with the incoming commander assuming responsibility for the mission. This changeover process may affect conditions under which the mission will continue.

Redeployment Considerations

5-66. Although intended for a direct handoff between U.S. units, commanders must make specific considerations when making a handoff to a multinational force. The considerations include mission, enemy, terrain and weather, troops and support available, time available, and civil considerations. For units relieved of a function by a government agency, procedures typically entail longer handoff times and more complex coordination. However, the other areas of consideration still apply and may in fact be a greater issue for an agency. Outgoing units that have past, present, or future projects planned with agencies prepare to transfer these projects to responsible agents in the incoming unit.

Mission

5-67. The outgoing unit must prepare a detailed study of their unit's mission statement and detail the current and implied mission tasks. The mission may also require an incoming unit with additional skill sets, such as advanced special operations, near real-time connectivity, civil affairs functional specialists, or complex media production ability. Knowing the mission, commander's concept of the mission, commander's critical information requirement, PIRs, and information requirements helps the incoming unit understand the mission. After a complete in-depth study of the operational area, the outgoing unit should help the incoming staff complete the handoff in a manner that allows for continued, uninterrupted mission accomplishment. The changeover must not allow the adversary to gain any operational advantages.

Enemy

5-68. The outgoing unit must provide the latest available intelligence on all threats that affect the mission. This intelligence includes comprehensive data on terrorists and terrorist-related incidents over the previous several months. In addition to the normal intelligence provided to the incoming unit on a regular basis, the situation may call for a liaison from the outgoing unit. Operations security is critical to prevent the enemy from discovering the impending handoff and then exploiting the fluidity of the change and the concentration of U.S. or multinational forces.

5-69. The in-country unit provides continuous information updates to the incoming unit. PIRs and information requirements were established for the original mission along with operational, strategic, and tactical information. The incoming unit must become familiar with the ongoing and upcoming mission PIRs and information requirements along with their links to planned nonlethal effects.

Troops and Support Available

5-70. To the incoming unit, learning about the friendly forces is as important as knowing the enemy situation. The unit must be familiarized with the command and control structure with which it will work daily. The incoming unit must know all friendly units in adjacent operational environments. It also must be aware of joint and multinational units and the capabilities of their mission support base. The incoming unit must also be aware of other operations, units, and their capabilities.

5-71. Mission handoff is the time for the outgoing unit leadership to introduce the incoming unit leadership to leaders of FSF. This practice helps prevent breaks in intelligence networks already developed by the outgoing unit and reduces the time the incoming unit needs to build trust. The outgoing unit supports the incoming unit's plans to prepare for the transition in counterpart relations. However, potential or anticipated friction between a foreign security force and the incoming unit may cause the relief to occur more slowly. Therefore, incoming and outgoing units need time for overlap to allow for in-country, face-to-face contact with their counterparts before the mission handoff. If possible, outgoing unit members should provide biographical data on their counterparts to include photographs. This information allows unit members to become familiar with their counterparts before deployment and to determine which advisor techniques may need more emphasis. Execution of the mission must continue within the capabilities of the incoming unit, the foreign unit, and the available supporting assets.

Time

5-72. The depth and dispersion of units and the number of operations conducted determines the time required to exchange units. Ideally, an overlap occurs to allow the incoming unit to become familiar with the operational environment and to establish rapport with their foreign counterparts. However, the handoff must occur as quickly as possible. The longer it takes, the more vulnerable become personnel in the operational environment. However, both outgoing and incoming units should not sacrifice continued and uninterrupted execution of ongoing operations for speed. The incoming unit needs enough time to observe training techniques and procedures and to conduct debriefing on lessons learned. Outgoing unit personnel need sufficient time to put incoming unit personnel in contact with agency counterparts.

Civil Considerations

5-73. The outgoing unit provides an in-depth area study, giving close attention to local problems. Often sister units can provide general demographic data that can be expanded upon for unit-specific needs. Popular support for U.S. activities occurring in the operational environment may directly influence changes in the mission statement. The outgoing unit provides this critical information and describes all completed and progressing civic action projects. The incoming unit considers how the host-nation government functions and international civilian or government agencies are involved in or influence the situation in its operational environment.

POSTDEPLOYMENT

5-74. The unit commander debriefs to provide an overview of the mission and all relevant information subsets. Topics can range from military geography, political parties, and military forces to insurgents, security forces, and ongoing agency operations. Concerns specific to a foreign security force and relevant in one SFA operation may not be relevant in another operation. The commander should codify and debrief any information subset that may affect mission success.

POSTMISSION DEBRIEFING

5-75. Redeployment is not the end of the mission. Upon arrival at the redeployment location, the unit undergoes an extensive debriefing. The intelligence staff section officer typically organizes and conducts the debriefing. The intelligence officer coordinates with higher-level intelligence organizations to take part in the returning unit's debriefing, particularly if other organizations tasked the unit to obtain information. All deployed personnel, to include attachments, must be available for the debriefing.

DOCUMENTATION

5-76. After the debriefing, the unit commander and staff prepare two documents—an after action review and a report of lessons learned.

5-77. The after action review states the who, what, when, where, and how of the operation. It is a permanent record of the major activities of the unit from receipt of mission to debriefing. As such, it is an extremely important template on which past missions may be compared and future missions planned. Within 48 hours of being debriefed, the unit normally submits an after action review through command channels to the higher command. The intelligence and operations officers at each echelon keep copies of unit after action reviews.

5-78. Shortly after completion of the after action review, or simultaneously with its submission, the unit submits a report of lessons learned. This report documents the commander and staff's reflection of the operation and recommendations for the future. Units often prepare the lessons according to the six warfighting functions: movement and maneuver, intelligence, fire support, sustainment, command and control, and protection. This method addresses what worked and did not work during the operation, why it did or did not work, and what changes or substitutions are needed for existing tactics, techniques, and procedures in the unit.

5-79. As a minimum, the lessons learned report is forwarded to appropriate Service- and joint-level lessons learned agencies to allow integration into subsequent unit training and leader education. These agencies might include the Center for Army Lessons Learned and Joint Center for Operational Analysis.

Chapter 6
Sustainment

A brigade combat team may be required to execute operations across the spectrum of conflict. Hence, providing security force assistance (SFA) may require the sustainment staff to analyze the mission to determine support requirements and available capabilities. Since brigade combat teams conducting SFA may be responsible for a large operational area, an accurate and thorough assessment of U.S. assets as well as the foreign security force's assets is critical to mission success. The sustainment planner understands that though the mission can change, methods by which logistics are distributed within a theater of operations do not. Army sustainment enables the commander to execute a mission and sustain the force through full spectrum operations. Sustainment assets are allocated based on those requirements; no standard arrangement fits all situations.

As discussed in chapter 2, one of the principles of SFA is to ensure long-term sustainment. This principle dictates that all policies, strategies, and programs must be durable, consistent, and sustainable by foreign security forces (FSF) and their government once U.S. forces have completed assistance. This can be accomplished by developing the foreign security force's capability and capacity for asset production, management, performance, and maintenance.

SUSTAINMENT IN FOREIGN SECURITY FORCES CAPACITY

6-1. Sustainment is the provision of the logistics, personnel services, and medical support necessary to maintain operations until mission accomplishment. The endurance of SFA forces is primarily a function of their sustainment. Sustainment determines the depth to which the forces can conduct decisive operations, allowing the commander to seize, retain, and exploit the initiative. Endurance is the ability to employ combat power anywhere for protracted periods. Endurance stems from the ability to generate, protect, and sustain a force. It involves anticipating requirements and making the most effective, efficient use of available resources. Sustainment also enables strategic and operational reach. SFA forces require strategic and operational reach to deploy and immediately conduct operations anywhere with little or no advanced notice. FM 4-0 discusses sustainment in detail.

CORE LOGISTIC CAPABILITIES

6-2. Similar to any operation, the core logistic capabilities in SFA provide a framework to facilitate integrated decisionmaking, synchronize and allocate resources, and optimize logistic processes. The challenges associated with support cut across all the core logistic capabilities: supply, maintenance, deployment and distribution, engineering, medical support, logistic services, and operational contract support.

Supply

6-3. Supply operations include identifying requirements; selecting supply sources; scheduling deliveries; receiving, verifying, and transferring products; and authorizing supplier payments. To provide responsive supply operations in SFA to both the advising organization and FSF, logistic planning requires a collaborative environment that includes operations planning, maintenance operations, and the distribution system.

Maintenance

6-4. In SFA, maintenance operations aim to reduce repair cycle times by replacing components, modules, and assemblies as far forward as possible as well as maximizing reliance on parts distribution, visibility, and replacement. Advisors and their foreign counterparts require maintenance capacity for their freedom of action. Force movement and maneuver depends on sufficient readiness levels of transportation equipment and weapons systems. In SFA, maintenance encompasses both intermediate and organizational levels and consists of shop maintenance supporting the supply chain, transportation assets, and weapon systems.

Deployment and Distribution

6-5. The deployment and distribution capability moves forces and logistic support on time by meeting the required delivery date and providing time-definite delivery. By sharing critical information, units create unity of effort among diverse distribution organizations. This unity of effort enhances foreign support efforts and provides end-to-end support to satisfy deployment execution and sustainment operations for advisors and host-nation forces.

Engineering

6-6. Engineering provides capabilities to maximize freedom of action. These include capabilities and activities that support the maneuver of land combat forces; modify, maintain, or protect the physical environment; and contribute to a clear understanding of the physical environment by providing geospatial information and services to commanders and staffs.

Medical Support

6-7. A component of Army health systems oversee operational management of the health service support and force health protection missions for training, predeployment, deployment, and postdeployment operations. Army health system includes all mission support services performed, provided, or arranged by Army Medical Department to support health service support and force health protection mission requirements for civilian and military joint and multinational forces. For more information on Army health systems, see FM 4-02.

Health Service Support

6-8. All support and services performed, provided, and arranged by Army Medical Department promote, improve, conserve, or restore the mental and physical well-being of Soldiers and personnel in other Services, agencies, and organizations as directed. This includes casualty care, medical evacuation, and medical logistics. The health service support mission is a function of the sustainment warfighting function.

Force Health Protection

6-9. Force health protection measures promote, improve, or conserve the mental and physical well-being of Soldiers. These measures enable a healthy and fit force, prevent injury and illness, and protect the force from health hazards. They include some prevention aspects of Army Medical Department functions, combat and operational stress control, dental services, and laboratory services. The force health protection mission is a function of the protection warfighting function.

Logistic Services

6-10. Logistic services comprise the support capabilities that collectively enable the United States to rapidly provide global sustainment for U.S. forces. Logistic services include many disparate activities that are highly scalable capabilities. Included in this area are food, water and ice, base camp, and hygiene services.

Contract Support

6-11. Operational contract support provides acquisition functions to obtain materials, services, and construction of facilities for U.S. and host-nation forces in the local operating areas. The Department of Defense increasingly relies on contractors to perform many functions and tasks and then provides technical oversight. The expeditionary nature of SFA creates an even greater need for this capability.

PRINCIPLES OF SUSTAINMENT

6-12. The principles of sustainment are critical to maintaining combat power, maintaining strategic and operational reach, and providing Army forces with endurance. The principles of sustainment are anticipation, responsiveness, simplicity, economy, survivability, continuity, improvisation, and integration. While these principles are independent, they are also interrelated when used in planning and executing operations. For more information on sustainment principles, see FM 4-0.

MISSIONS

6-13. All brigade support battalions have the same general capability to manage distribution and sustainment operations. The brigade support battalion provides support within an assigned operational area. It is force tailored and task-organized and uses subordinate companies, platoons, and teams to perform specific functions. In the sustainment role, the brigade support battalion is primarily concerned with the continuous management and distribution of stocks and allocation of maintenance in the operational area. Such assistance provides operational reach to maneuver commanders.

6-14. The brigade support battalion contains the following sustainment capabilities:

- Provides supplies, field services, field maintenance, recovery, and field feeding for itself and its assigned subordinates.
- Plans and conducts base and base cluster self-defense; defends against level I threats, assists in destruction of level II threats, and escapes or evades against level III threats.
- Has ability to operate as part of an Army or joint force.
- Coordinates host-nation support.
- Has ability to deploy an advance party to support early entry operations.
- Provides opening, distribution, and sustainment management information and advice to commanders and staff within its operational area.
- Exercises technical supervision over operations for all assigned units.
- Provides limited materiel management for internal stocks—classes I, II, III, IV, V, VII, and IX and maintenance management of internal assets.
- Provides a liaison team to augment other headquarters as necessary.
- Manages and maintains the its property records.
- Maintains data in support of Army equipment status reporting database and Army equipment status reporting system.
- Appoints contracting officer representatives to oversee contractor performance, certify receipt of services, and act as liaisons between the requiring activity and the contracting officer.

6-15. A brigade combat team often has attached companies such as an engineering company, a military intelligence company, an antitank company, or a network support company. Similar to the brigade headquarters and headquarters company, these attached companies are supported by the brigade support battalion regardless of where they are located on the battlefield. If one or part of these companies is task-organized to a maneuver battalion, it remains under the support of the brigade support battalion.

Critical Considerations in Building Host-Nation Sustainment Capacity

6-16. Effective SFA planning begins with accurately assessing and understanding U.S. policy objectives, determining the end state, and identifying resources available. Some possible logistic planning considerations—

- Have a holistic perspective.
- Address the causes, not the symptoms.
- Understand the desired end state.
- Understand the level of FSF senior leader support.

6-17. Logistic planning during SFA uses a holistic perspective. Operational environments can range from peacetime cooperation to counterinsurgency, stabilization to reconstruction. SFA forces determine the foreign security force's logistic and reconstruction requirements based on their internal defense and development strategy; they develop a sustainment end state to support the plan. Rebuilding the host nation's infrastructure and industrial capacity not only provides long-term job opportunities, but also builds trust and legitimacy in FSF, the host-nation government, and partner organizations.

6-18. Logistic planning during SFA addresses the causes, not the symptoms. Operational logistic support problems are the symptoms, not the causes. They result from problems that exist in the supporting systems or in the society's industrial capacity and economic strength. Teaching unit-level inventory management alone will not result in any sustainment improvement.

6-19. Logistic planning during SFA understands the desired end state. Depending on the U.S. strategic interest in a host nation and its security forces, levels of engagement can vary significantly to achieve the desired end state. Sustainment capacity-building activities can be force tailored and scaled to assist in any of the sustainment capability areas.

6-20. Logistic planning during SFA understands the level of FSF senior leader support. These leaders develop and set the policies for resource allocation. They support the various development projects and skills enhancement assistance activities. These senior-level advisors must fully understand the problems identified and an assistance strategy developed by SFA planners. From the beginning, SFA planners work with their counterparts to decide the plans and strategy to promote the vision and ensure continued effort.

6-21. SFA depends on capable logistic support for maneuver and sustainment. Without an improvement in sustainment capacity, FSF cannot significantly improve regardless of how much effort is devoted to advising and training. When planning for sustainment capacity building, planners thoroughly understand the logistic systems to identify causes that lead to functional deficiencies. They must have a holistic perspective and understand the various goals and resources required to formulate an operational approach that effectively and efficiently achieves U.S. objectives.

ASSET MANAGEMENT AND PERFORMANCE

6-22. One tool at the disposal of the commander and planners during an evaluation of foreign infrastructure and industrial capacity is called an asset management plan. This plan, commonly employed in the industrial community, manages an organization's infrastructure and assets to deliver a standard of service. An asset management plan provides information to FSF so they can analyze the capacity, condition, and current and future requirements of their facilities and infrastructure. Physical development should include facilities, transportation networks, area development, utilities, and other infrastructure components.

6-23. A description of the perceived problems of current foreign facilities, or problems created by missing assets, is also important. The description includes how the advisor intends to address deficiencies in foreign facilities and infrastructure. Each organization needs to determine each facility's function and value to their organizational mission by analyzing the current asset performance, costs, benefits, and improvements.

Current Asset Performance

6-24. Asset performance is the most difficult portion of the asset management plan process to capture while assessing conditions of facilities and infrastructure. While historical records may exist, combat operations and possible instability of the host-nation government likely render these records irrelevant. Remaining structures, when measured against U.S. construction standards (designed to protect occupants from man-made or natural disasters), often are deficient, even if untouched by hostilities. Most existing facilities require significant modification to be considered for future use. However, FSF may use them in the interim until new facilities are completed.

6-25. Asset performance may also use gap analysis. Gap analysis defines the present state, the desired end state, and the gap between them. Once a gap has been identified, the commander and staff can develop courses of action to rectify or bridge the gap. Gap analysis alone is not adequate for all problem sets, as goals may evolve and emerge during the course of problem solving. Most problems have many alternative solutions.

Costs

6-26. Establishing the short, medium, and long-term costs for facilities and infrastructure, from construction to control, is a critical step in prioritizing which facilities get built or refurbished first, and which get support at all. Units develop a forward-looking cost-profile for operating, maintaining, refurbishing, and replacing assets. This profile explains to civilian and military officials the implications and continued costs to maintain assets. SFA requires a plan for U.S. forces to hand off assets to FSF.

6-27. In constructing these models, the ideal cost-profile extends to cover the life of the longest-lived asset in the system. By doing so, the profile estimates the whole-life cost as well as determines the average annual costs. Furthermore, this cost evaluation may need to be revisited and updated annually, forming the basis for an annual bid for funding.

Benefits

6-28. Assets should provide some measure of benefit to U.S. and foreign security forces that can be measured or explained during and after SFA. This usually involves translating asset costs and projected performance into a monetary figure. Other benefits may be social or environmental, which may be difficult to quantify in monetary terms. Regardless, some attempt to record all the relevant benefits is important; the more quantitative, the better.

6-29. A benefit evaluation is vital to demonstrating that the ongoing cost of an asset system is reasonable given the scale of the benefits. The expenditure incurred by asset management needs to be justified in some way.

Improvements

6-30. How assets of FSF perform may be improved by acquisition, enhancement, or some other means. An improvement may address the potential reduction of performance and its effect on the rest of the asset system. The improvements represent a potential change to the performance and are typically managed as a project. Such management comes with an appraisal of the additional expenditure, comparing different options and selecting a preferred option based on the commander's guidance and host-nation requirements.

OPERATIONAL CONTRACT SUPPORT

6-31. Contracting is a key source of support for deployed forces across the spectrum of conflict. Because of the importance and unique challenges of operational contract support, the commander and staff need to fully understand their role in planning for and managing contracted support in the operational area. This includes understanding the types of contracted support, contract management, and contract close out.

TYPES OF CONTRACT SUPPORT

6-32. Current doctrine describes three broad types of contracted support: theater of operations support, external support, and systems support.

Theater of Operations Support

6-33. Contracts for theater of operations support provide support to deployed forces under prearranged contracts or contracts awarded from the mission area. They do so by contracting officers under the command and control of the combat sustainment support battalion. Contractors for theater of operations support acquire goods, services, and minor construction support, usually from local commercial sources, to meet the immediate needs of operational commanders. Theater of operations support contracts are the type of contract typically associated with contingency contracting. Sustainment brigades often are the requiring activity for theater of operations support contract actions.

External Support

6-34. Contracts for external support provide a variety of support to deployed forces. External support contracts may be prearranged contracts or contracts awarded during the contingency. These contracts support the mission and may include a mix of personnel including U.S. citizens, third-country nationals, and host-nation subcontractor employees. The largest and most commonly used external support contract is the logistics civilian augmentation program. This Army program is commonly used to provide life support, transportation support, and other support functions to deployed Army and joint forces.

System Support

6-35. System support contracts are prearranged and include, but are not limited to, newly fielded weapons systems, command and control infrastructure, and communications equipment. System support contractors, made up mostly of U.S. citizens, provide support in garrison and may deploy with the force to both training and real-world operations. The unit does not normally have a significant role to play in planning for or coordinating system support contracts other than coordinating and executing support of system support contract-related personnel.

CONTRACT MANAGEMENT

6-36. Contract management requires nominating and tracking contracting officer representatives for every service contract. It will also nominate a receiving official for all supply contracts. To ensure contractors provide the service or item per the contract, they need a quality contracting officer representative and official support. SFA units must identify and coordinate funding requirements in advance with the supporting resource management staff element to secure the correct amount and type of contract funding associated with the mission.

CONTRACT CLOSE OUT

6-37. In long-term operations like SFA, the advising unit ensures direct coordination and transfer of information related to operational contract support before mission handoff. Additionally, when the advance party arrives in the operational area, a designated Soldier actively seeks out current information on local contract support capabilities, policies, and procedures. This individual coordinates the formal handover of existing contract management responsibilities from the redeploying unit. This Soldier must know when recurring service contracts end since it takes 30 to 60 days to obtain funding approval. If the unit waits until the contract is about to expire before requesting additional funds, the sustainment brigade could lose the contracted service until funds are available.

EXTERNAL AGENCIES SUPPORT MANAGEMENT

6-38. Some external agencies help support the management of FSF. External support comes from interagencies, the host nation, and multinational organizations.

INTERAGENCY SUPPORT

6-39. The Department of Defense performs both supported and supporting roles with other agencies. It often is the lead agency for logistic activities in SFA. When tasked to provide military support to civil agencies, Army forces perform a supporting role. Sustainment forces may be tasked to support the U.S. ambassador that, as the President's representative, serves as senior U.S. Government officer in the country, or may employ resources of other U.S. Government agencies or private firms. Whether supported or supporting, close coordination is the key to efficient and effective interagency operations. In the absence of an American Embassy, or during operations in austere environments or remote locations, the advising unit may be the only organization able to provide supplies, services (including force protection), and life support. More typically, during the early stages of the advising unit deployment, the unit may find the American Embassy to be the primary source of supplies, services, and life support.

6-40. Nongovernmental organizations do not operate in either the military or governmental hierarchy. Their relationship with an advisor organization is neither supported nor supporting. An associate or partnership relationship exists between military forces and engaged nongovernmental organizations, ideally focused on unity of effort. If formed, the focal point where U.S. forces provide coordinated support to nongovernmental organizations would be the civil-military operations center of a joint task force headquarters.

HOST-NATION SUPPORT

6-41. Host-nation support and local procurement may provide sustainment, operational support, and tactical support. Host-nation support agreements fulfilling the command requirements for support need to be prenegotiated. Such support arrangements must be integrated into the distribution plan and coordinated with joint, allied, and multinational partners to prevent competition for resources and ensure high priority requirements are met. Host-nation support may include functional or area support and may use host-nation facilities, government agencies, civilians, or military units. Preestablished arrangements for host-nation support can reduce the requirement for early deployment of U.S. assets and can offset requirements for early strategic lift by reducing requirements for moving resources to the theater of operations.

MULTINATIONAL SUPPORT

6-42. Multinational support may consist of support provided from one multinational partner to another. One or more of the following organizational and management options facilitates multinational support:

- National support elements provide national support.
- Individual acquisition and cross-servicing agreements provide limited support.
- A lead nation provides specific support to other contributing nation forces.
- A role-specialist nation provides a specific common supply item or service.
- A multinational integrated sustainment unit provides limited common supply and support.
- A multinational joint sustainment center manages common-user logistic support.

6-43. In all cases, the multinational force commander directs specific multinational support within the applicable laws and regulations of the host nation. When operating within a formal alliance, the advising unit executes support following applicable standardization agreements.

FACILITATING THE REDEPLOYMENT PROCESS

6-44. Sustainment assets facilitate the redeployment and the retrograde process through its command and control of the distribution system. Many of the procedures are used to deploy forces, draw pre-positioned stocks, conduct reception staging and onward movement, and distribute supplies within the theater of operations apply to the redeployment process. Two factors in particular complicate redeployment operations: supporting operations and meeting requirements.

6-45. The same elements that operate and manage the theater of operations distribution system during deployment and sustaining operations perform similar roles during redeployment. When redeployment,

deployment, and sustaining operations occur simultaneously, the theater sustainment command may find it necessary to rebalance its forces or change the missions of subordinate units to effectively support ongoing operations and redeployment.

6-46. Requirements vary widely depending on the nature and scale of redeployment operations, infrastructure, and other factors. For example, redeployment operations could range from personnel only to entire units and their equipment. Depending on the political and military strategy, unit rotations may occur while decisive operations continue unabated or during operational transitions. Key considerations include, but are not limited to, several characteristics. These characteristics include the size of the force redeploying and deploying; infrastructure requirements and limitations; security requirements; traffic circulation; staging areas; distribution system capacity; quantity of supplies and materials to be redistributed; the amount and kinds of materials that require disposal agricultural inspections; and establishing accountability of retrograde cargo. The challenge is effective coordination and synchronization, vertically and horizontally, to ensure responsive simultaneous support to not only ongoing distributed operations, but also redeployment.

Chapter 7
The Advisor

Advisors are the most prominent group of U.S. personnel that serve with foreign security forces (FSF). They live, work, and fight with their FSF. The relationship between advisors and FSF is vital. Advisors are not liaison officers, nor do they command FSF units.

ROLES OF THE ADVISOR

7-1. The military advisor has three roles involving different responsibilities. First and foremost, advisors are members of a U.S. military organization with a well-defined chain of command and familiar responsibilities. Second, advisors embed themselves with their counterparts. Third, advisors are interpreters and communicators between U.S. forces and their foreign counterparts.

7-2. As members of military organizations, advisors receive and execute the orders of superiors. These orders may conflict with the orders their counterparts receive. Among other duties, advisors must act unobtrusively, but nonetheless positively, as inspector general—often observing, evaluating, and reporting on the performance of counterparts and their assigned unit.

7-3. Secondly, advisors live, eat, and work with the officers and men of their host units. Often, advisors soon regard themselves as one of them. The sharing of common hardships and dangers forges potent emotional ties. The success and good name of their units become matters of prime and personal importance to the advisor.

7-4. Finally, advisors are interpreters and communicators between U.S. superiors and foreign counterparts. Advisors must introduce and explain one to the other; they help resolve the myriad of problems, misunderstandings, and suspicions which arise in any human organization, particularly when people of starkly different cultures approach difficult tasks together. Advisors with quick and easy access to influential counterparts can sometimes be the best possible means of communicating.

7-5. To be effective, advisors obviously must gain their counterparts' trust and confidence. This relationship, however, is only a prelude to the advisor's major objective: inspiring and influencing a counterpart to effective action. In pursuing this goal—constantly, relentlessly, and forcefully, yet patiently, persuasively, and diplomatically—advisors must recognize conditions which can benefit or handicap their cause.

CONSIDERATIONS OF THE ADVISOR

7-6. An advisor follows ten considerations when serving with FSF.

BY, WITH, AND THROUGH

7-7. All planned operations of a combat advisory mission must be conducted by, with, and through the FSF. Not counting immediate action battle drill responses, the mark of an effective advisory effort is the amount of personal involvement the FSF take in their own operations. Civilians must see that they are secured by their own security forces to promote the legitimacy of the host nation and their capacity and to build trust and confidence. In a word, combat advisors are shadows whose presence is felt at all times.

Empathy Leads to Competency

7-8. Empathy can be defined as identification with and understanding of another's situation, feelings, and motives. This is tough for experienced U.S. leaders and often harder to explain, but it is the key to the success of an advisory mission. Truly understanding other human beings and where they come from allows for honest relationships to develop. In most cultures, the place to begin understanding another person's feelings and experiences is by understanding the other person's narrative. The narrative is a collective group's identity as an interpretation of both ancient history and recent collective experiences. Delving into the narrative, understanding it and how it affects people, is the beginning of empathy.

Success Is Built on Personal Relationships

7-9. Developing a sense of empathy allows advisors to begin nurturing relationships with those they are tasked to advise. No amount of resources and firepower can compensate for the lack of a relationship between advisors and their counterparts. It must be honest, genuine, and heartfelt. Mutual respect, trust, and understanding create success. Both parties rely on each other for mission accomplishment and often for survival. This relationship is likely to be challenged on numerous occasions; only one built on mutual trust can survive and ensure mission success. Furthermore, advisors learn to pick their battles. Camaraderie and rapport with FSF can vanish in an instant if one does not prioritize and define critical lines in the relationship.

Advisors Are Between U.S. and Foreign Security Forces

7-10. Living and fighting with FSF allows for the development of true bonds, empathy, and trust. These desired results have a price to pay. When addressing their own Army, advisors soon realize that they are also not one of them, increasing their level of frustration. Advisors are often alone, navigating between two military systems and two cultures, never quite fitting in with either one.

Advisors Will Never Win, Nor Should They

7-11. Often advisors find it difficult to understand what victory looks like in this environment. Progress tends to occur at a glacial pace and cannot usually be tracked on a day-to-day basis. The advisor attaining a tactical objective does not achieve success; success is achieved by the foreign security force achieving the objective. Advisors find it difficult to satisfy their own units, and they never fully satisfy the demands of their FSF. They are figuratively and literally caught in the middle. Advising is the art of striving to make a win-win situation for all parties. The advisor and foreign security force relationship is one in which the advisor removes the obstacles in the way. Only a plan by FSF will succeed—and it will only succeed if it is, in fact, their plan.

Advisors Are Not Commanders

7-12. Advisors do not command FSF personnel or units. Advisors provide advice, training, and access to coalition resources to FSF. They are not intended to lead FSF in combat; they are responsible for the command and control of their own team of advisors. However, they can and should influence the commander of the foreign unit.

Advisors Are Honest Brokers

7-13. Advisors are advocates for FSF with conventional forces. They transparently assess capabilities and reveal limitations of foreign units to the FSF's and advisors' higher command. They also provide access to U.S. resources—from close air support to medical evacuation—to FSF they advise.

Advisors Live with Shades of Gray

7-14. Black and white is not available to the advisor. Caught between two cultures, systems, and narratives, the advisor works within a gray area. Unlike operations in a conventional unit, many of the checks and balances are not present. Advisors often find themselves isolated with great autonomy and no supervision. These conditions present moral and ethical dilemmas daily. Advisors need to be comfortable in this environment. Often they lose sleep at night and question their actions and those of their FSF. Thus, this mission calls for Soldiers of enormous character, moral courage, and intellect.

Talent Is Everything, But Advisors Must Understand Rank

7-15. Military forces around the world approach rank and prestige in completely different ways. In some cultures, rank equals experience; in others, it is nobility or education. Often many militaries have rewarded loyalty with promotions, which results in rank-heavy armies. Some FSF recognize talent, while others recognize rank. Advisors likely advise counterparts much more senior in rank. They must understand that rank on the uniform is important to many armies.

Make Do

7-16. Advisors will never have everything they believe they need to succeed. However, their creativity often distinguishes between success and failure. Austerity and economy of force can fail to meet the prevalent expectations that FSF have of a U.S. unit in regards to resources. It is up to the advisor team to close the expectation gap by its creativity and imagination. Scrounging, bartering, and negotiating are daily activities of advisors. An enormous amount of energy must be devoted to these activities to achieve mission success and endear them to their counterparts.

PERSONALITY TRAITS OF THE ADVISOR

7-17. Not every Soldier is well suited to perform advisory functions; even those considered to be the best and most experienced have failed at being an advisor. Effective advisors are only the most capable individuals. Advisors are Soldiers known to take the initiative and who set the standards for others; however, they are also patient and personable enough to work effectively with FSF. Recognizing that not all Soldiers are capable of performing as advisors, leaders should immediately remove advisors who do not exhibit these qualities.

7-18. Because advisors operate in very subjective environments, it is difficult to establish objective criteria by which to assess potential advisors. However, research and experience indicate that several personality traits greatly enhance the advisor's ability to adapt and thrive in a foreign culture. These traits include—

- Tolerance for ambiguity.
- Realistic when setting goals and tasks.
- Open-mindedness.
- Ability to withhold judgment.
- Empathy.
- Communicativeness.
- Flexibility.
- Curiosity.
- Warmth in human relations.
- Motivation of self and others.
- Self-reliance.
- Strong sense of self.
- Tolerance for differences.
- Perceptiveness.
- Ability to accept and learn from failure.
- Sense of humor.

7-19. Of the traits listed above, no single trait is paramount. Advisor selection and training programs seek to develop an understanding of the contributions of each of them.

SKILLS OF THE ADVISOR

7-20. Advisor-specific skills fall into two subcategories: enabling skills and developing skills. Enabling skills build on the individual and collective skills of advisors. Developing skills are the main advisor tasks of teaching, coaching, and advising. Advisors use these skills to develop the capabilities of the foreign security force.

ENABLING SKILLS

7-21. Advisors use enabling skills to work individually and in a group. Enabling skills help advisors communicate across cultures, build rapport, influence, and negotiate. Successfully employing these enabling skills sets the conditions for advisors to move forward with their mission.

Individual Skills

7-22. Individual skills help advisors properly perform their mission. Initially advisors should be selected based on subject matter expertise and not just rank. However, tactical proficiency does not necessarily equate to proficiency as an advisor. Modern-day military advisors must be able to teach and advise FSF. To do this, they must be knowledgeable and proficient in procedures and able to impart this knowledge effectively. They may also be able to advise FSF in the areas of intelligence, communications, operations, and logistics. Some advisors may be required to call in U.S. supporting arms, principally artillery and air support, and coordinate both air and ground casualty evacuation. Finally, advisors should be proficient in basic combat skills. These common skills, warrior skills, and survival skills are often required by advisors. Since these skills are critical to an advisor due to the isolated and independent nature of the mission, advisors should refresh them during predeployment training.

Collective Skills

7-23. As a group of advisors form a team, they must practice their warrior tasks, organize to cover the functions associated with any team, and familiarize its members with their duties and responsibilities. As with individual advisor skills, some are refined before deploying and others are learned in country. Figure 7-1 depicts this graphically.

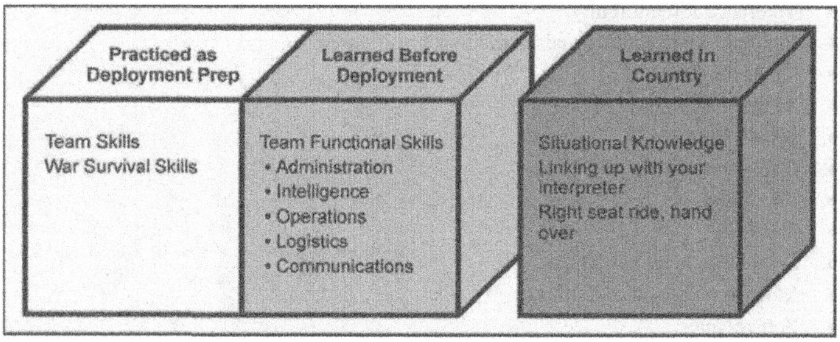

Figure 7-1. Collective advisor skills

DEVELOPING SKILLS

7-24. Developing skills—teaching, coaching, and advising—are required for every advisor. The advisor will make an initial assessment to determine the foreign unit's proficiency to determine the appropriate developing skill with which to start. Figure 7-2 illustrates the appropriate level of advice that corresponds to the foreign unit's capabilities.

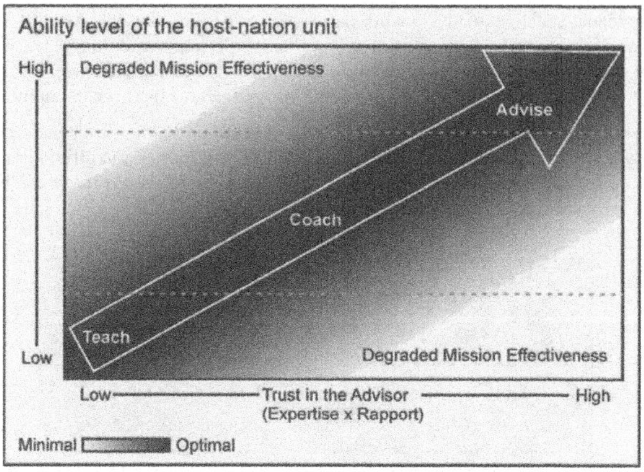

Figure 7-2. Appropriate advising level

7-25. Advisor teams have limited positional power and depend on personal influence to positively affect FSF efforts through teaching, coaching, and advising. Advisor team members often work with people of higher rank or grade than themselves. Advisors should remember the foreign unit's experience and capabilities and carefully choose opportunities to inject or impart knowledge. Foreign units are most receptive to advisor teams that teach unobtrusively. Foreign units most value those who are subtle in their teaching, coaching, and advising. Those advisors who master the ability to create a climate in which FSF personnel feel they are teaching themselves often prove the most effective.

Teaching

7-26. Teaching includes training and education. Methods of teaching can include classroom lectures, seminars, hands-on training, training exercises, and simulations.

Coaching

7-27. Coaching relies on guiding to bring out and enhance capabilities already present. Coaching refers to the function of helping someone through a set of tasks. Those being coached may or may not have appreciated their potential. The coach helps them understand their current level of performance and instructs them how to reach the next level of knowledge and skill. Coaching requires identifying short- and long-term goals and devising a plan to achieve those goals. The coach and the person being coached discuss strengths, weaknesses, and a course of action to sustain or improve the goals.

Advising

7-28. Advising is a combat multiplier that boosts supported unit capability. Advisors have experience in particular areas in which they are advising, but are not required to have similar backgrounds. Advisor relationships are not based on superior to subordinate relationships.

7-29. Advisors provide an expert opinion, advice, and counsel by focusing on both personal development (interpersonal and communication skills) and professional development (technical and tactical knowledge). Advising develops mutual trust and respect.

Advisor Versus Trainer

7-30. Advisors teach, coach, and advise work in a cycle. As FSF master one skill, the advisor can move on to other skills and initiate the process for the new skills. If FSF require additional teaching or coaching, the advisor can take them to that part of the cycle. Eventually, the teaching and coaching should decrease, and most of the advisor's time will be spent as an advisor providing an expert second opinion.

7-31. Ultimately, the question of what is the difference between being a trainer or an advisor will arise. Figure 7-3 shows that every advisor will have to be able to be a trainer, but all trainers are not expected to perform the more sophisticated task of advising. Hence, competent junior officers and enlisted Soldiers can be trainers, but only more experienced individuals can be advisors.

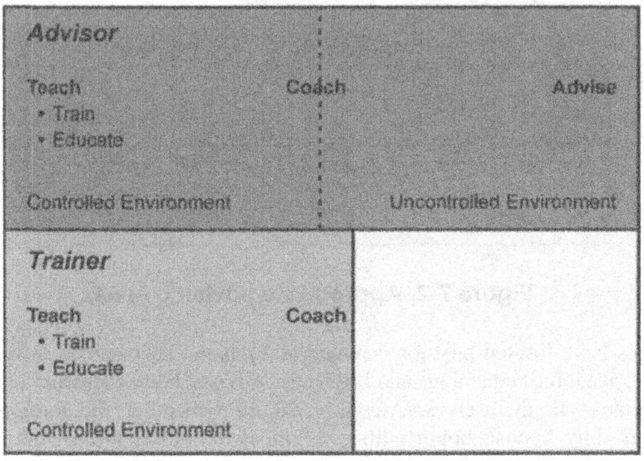

Figure 7-3. Advisor versus trainer

Chapter 8

Culture and Communication

As an advisor, understanding the host-nation population is a crucial element of premission planning and the development of foreign security forces (FSF). Prior knowledge of sociocultural differences aids in building effective relationships and prevents embarrassment, loss of rapport, and compromise of the mission. This chapter discusses three aspects of the sociocultural environment to provide a framework for analysis and comparison of foreign cultures for operational purposes. These aspects are society, culture, and communications.

SOCIETY

8-1. People who have a common culture, share a group identity, occupy a common territory, and are subject to the same political authority can be identified as a society. A society is not easily created or destroyed. Though they share the above factors, societies are not homogeneous nor static. A society usually has a dominant culture but can also have a vast number of secondary cultures, each influencing the other.

8-2. Advisors derive much of their effectiveness from their ability to understand and work with counterparts from a foreign society. Advisors must also consider societies outside their immediate operational environment whose actions, opinions, or political influence can affect the mission.

SOCIAL STRUCTURE

8-3. Society is composed of both social structure and culture. Social structure refers to the relations among groups of persons within a system of groups and is persistent over time. For the purposes of security force assistance, social structure includes groups, institutions, organizations, roles and statuses, and social norms. FM 3-24 contains a more in-depth discussion of social structure.

Groups

8-4. Within societies, groups are composed of two or more people that interact based on mutual interests and interrelated statuses and roles. These groups can be based on race, ethnicity, religion, or kinship. A racial group is defined by its members or by outside groups as different based on physical characteristics or descent. Race is a social category that has no biological basis. An ethnic group is a community whose learned cultural practices, language, history, ancestry, or religion distinguish them from others. Similar to racial groups, ethnic groups can be defined by their members or by outside groups. Religious groups may be subsets of larger ethnic groups and an ethnic group may contain members of different religions. Kinship groups are based on a link by blood relation. It may also be used to indicate a group of persons who live and work together to satisfy basic collective needs and goals. The role of a kinship group is generally seen to care for and educate offspring. This group provides for such needs as food, shelter, and clothing and prepares offspring to be functional members of the society in which they live.

Institutions

8-5. Institutions are the basic building blocks of societies. They are made up of groups that are organized to achieve a common task. Institutions are continuous through many generations and are built to continue to function even as individuals are replaced. An example of an institution from FM 3-24 is a military institution that brings together groups and individuals whose statuses and roles concern defense and security.

Organizations

8-6. Organizations are institutions that have a bounded membership with defined goals and established operations. They include fixed facilities from which they conduct business or operations and a set means of financial or logistic support. The four types of organizations are communicating, religious, economic, and social. Communicating organizations are designed to influence perceptions. Religious organizations are created to regulate societal norms, reaffirm beliefs and values, and provide social support. Economic organizations provide employment and assist in economic development. Social organizations provide support to groups in society, create social networks, and can influence ideologies. Organizations may belong to more than one category and exist to affect the behavior, values, and beliefs of society. As such, advisors must identify influential organizations both inside and outside of the FSF with which they are partnered.

Roles and Statuses

8-7. As members of a group, individuals interact based on assumed or given social positions. These social positions are referred to as statuses. Societies generally associate particular statuses with particular social groups (such as family, lineage, ethnicity, or religion). Each status inherently contains a set of expected behaviors known as a role, which includes how a person of that status should act, feel, and think. A status includes expectations about how others in a society should treat a person of that status.

Social Norms

8-8. Violation of a role derived from a given status results in social disapproval. The standard of conduct for social roles is a social norm. Norms may be either moral (incest prohibition, homicide prohibition) or customary (prayer before a meal, removing shoes before entering a house). Advisors must understand the roles, statuses, and social norms of groups they are advising. This knowledge will clarify the expected behavior of their FSF partners and their role as advisors.

Language

8-9. Language is a learned element of culture that is a system of symbols that people use to communicate with one another. Successful communication within a society requires an understanding of the social setting, appropriate behaviors towards different statuses and roles, and nonverbal cues. The languages used in the advisors' operational environment have a major impact on the advising mission. Languages must be identified to facilitate language training during predeployment and employment, as well as in the requisitioning of translators. Translators are critical for communicating with FSF and interacting with local citizens and leaders.

CULTURE

8-10. Social structure comprises the relationships among groups, institutions, and individuals in a society; in contrast, culture provides meaning to individuals within the society. Social structure can be thought of as a skeleton with culture being the muscle on the bones. The two are mutually dependent and reinforcing. A change in one results in a change in the other.

8-11. Culture influences an individual's range of action and ideas, including what to do and not do, how to do or not do it, and whom to do it with or not to do it with. It is neither a random jumble of customs nor a laundry list of accepted practices and taboos that an advisor should know. It influences judgments about what is right and wrong and what is important and unimportant. Cultural rules are flexible in practice; the customs and actions that an individual practices are based largely on that particular environment.

Cultural Characteristics

8-12. Culture is shaped by history, language, religion, and customs. Members of society use these shared beliefs, values, customs, and behaviors to cope with their world and with one another. It is learned behavior that is shared by members within a society. However, behaviors do not always define a culture.

Frequently, there are more telling cultural signs in the meaning attached to a behavior than in the actual behavior. Culture influences its members' identity, values, belief systems, and cultural forms. As such, it is arbitrary; advisors should make no assumptions regarding what a society considers right and wrong, good and bad. Culture is changeable through the social interactions between people and groups, both inside and outside of that society. TC 31-73 discusses cultural characteristics in great detail.

Identity

8-13. Each individual belongs to multiple groups through birth, assimilation, or achievement. Each group to which individuals belong influences their values and beliefs. Individuals consciously or unconsciously rank their identities into primary and secondary identities. Primary identities are those seen as core to their being; frequently these are national, racial, and religious. In contrast, secondary identities are more peripheral and may include such things as an individual's occupation or hobby.

Values

8-14. Values guide how individuals interact with the world around them. As deeply held beliefs about rightness, values shape how individuals think things should be. Values may never be articulated—in fact, they may not even be consciously held. At times, the beliefs of an individual may even contradict one another. Advisors should have some understanding of a counterpart's value system. Usually values are nonnegotiable. In working with a counterpart with conflicting values, advisors should persuade the individual that the advocated action or behavior really does not conflict with their values.

Belief Systems

8-15. Beliefs primarily take two forms: values (core) and opinions (peripheral). Values, or core beliefs, are not easily changed. Opinions, or peripheral beliefs, stem from thought and decisions. The totality of the identities, values, and beliefs that an individual holds compose that person's belief system.

Forms of Beliefs

8-16. Values are broad moral conclusions about the way to live life. These core beliefs are unstated, taken for granted, resistant to change, and not consciously considered. They reflect strong personal certainties drawn from faith, childhood environments, and life experiences. They lend shape and order to people's lives, helping them to decide if they settled on the right actions, friends, leaders, religions, and careers.

8-17. Opinions, or peripheral beliefs, are immediate conclusions about some aspect of the environment. They are usually quite specific and are open to debate, consciously considered, and easiest to change.

Types of Belief Systems

8-18. Belief systems involve cultural narratives and myths. The interpretation of these narratives and myths provide people with insight into how they should feel, think, and behave. The most prominent systems of beliefs tend to be those associated with religions and ideologies. However, any system of belief in which the interpretation of narratives impacts people's behavior (such as superstitions) can be a contributing component of a society's culture.

8-19. A belief system acts as a filter for new information. It is the lens through which people perceive the world. Beliefs of one group as to what is rational, normal, or true may appear strange, irrational, or illogical to outsiders. Understanding belief systems of various groups in an operational environment allows advisors to work with their foreign counterparts and influence civilians more effectively.

8-20. Belief systems play a fundamental role in the development of a society's values. Careful analysis of a region's belief systems is crucial to understanding the local value system. These systems influence religious doctrine, narratives, and myths and provide enormous insight into the culture of the community. Ultimately, the values which are most prized within a belief system are religious. These beliefs affect that culture's systems of social organization, learned behaviors, and relationship to the physical environment.

Religion as a Belief System

8-21. At its most basic level, religion is both the individual and communal expression of contact with the sacred. Notions of the sacred can vary widely in societies and may include deities and sacred items. In most cases, religions also involve symbols, rituals, and ethics or dogma. Religion can be both an individual practice and a group phenomenon.

8-22. Most societies have some variety of religion—typically a set of sacred beliefs and rituals that provide a common understanding of moral codes and proper conduct. Expressing a common religious understanding helps people to make sense of their world. Often religion provides explanations for human suffering, natural disasters, broken relationships, inequalities, and death. Some societies also have secular equivalents of religion. For instance, America is often remarked to have a civil or civic religion that can be seen in the images, symbols, ideas, and behavior of the political realm.

8-23. Because religion is such an integral part of culture, when advisors carefully prepare and analyze the mission, they must examine the religions and religious groups in the operational environment. Religious beliefs, leaders, and institutions are central to how many societies view the world. When planning any advising operation, planners consider how religion affects FSF and civilians.

Broad Impact of Belief Systems

8-24. In almost every society, followers of dominant religions have a strong societal impact. This influence may be exerted at varying degrees upon the individual citizen, sections of society, the economy, the military, and the political structure. TC 31-73 discusses these in detail.

THE IMPORTANCE OF UNDERSTANDING CULTURE

8-25. Advisors derive their effectiveness from their ability to understand and work with foreign counterparts from another culture. They must understand enough of their own culture and their counterpart's culture to accurately convey ideas, concepts, and purpose without causing counterproductive consequences.

8-26. Advisors must be aware of aspects of the local culture and history that influence behavior in their operational environment. Advisor team members must understand the reasons and motivations underlying personal interaction and must practice patience when working with their counterparts. Group norms guide individual behavior, and advisors must understand how individuals in a society tend to interact as members of a group, whether a race, ethnic, or kinship group.

8-27. Cultural understanding is not derived from demographic information provided to the military through country briefs prior to deployment. It is gained from studying, interacting, and understanding the people, religion, history, customs, and social and political structures within an area. For true understanding, it is necessary to live among the people, gradually understanding the subtleties and nuances of their culture.

CULTURAL EDUCATION

8-28. Culture influences a person's perception of the world. People from one society tend to view people from other societies through the prism of their own culture. Often, their perceptions are biased, are inaccurate, or involve assumptions of superiority. For example, Americans view their society and its culture very differently from the rest of the world. Likewise, as foreigners, advisors view local cultures through the prism of their own idealized culture. This creates a large gap between the culture the advisor observes and the host nation's real culture.

8-29. To narrow the gap of cultural understanding an advisor will encounter, a program of cultural learning should be undertaken. The end state of this cultural learning should be to gain the trust of foreign partners and civilians. Advisors will do so by demonstrating their understanding and sensitivity towards the host nation's culture and social structures.

8-30. An awareness of the cultural aspects of the operational environment can significantly enhance the effectiveness of an advisor. An extensive study program is best, but even a concise program is worthwhile if it provides advisors with the necessary cultural information. The best programs involve credible teachers who intimately understand the culture and have personal experience in the society. Training is dynamic and moves beyond the standard lecture and handout format. Small groups or a discussion panel work quite well. Cultural behavior and cultural familiarization handbooks can also be used. Since handbooks vary greatly in accuracy and quality, credible subject matter experts should be consulted to determine which documents to incorporate into the training. Table 8-1 discusses the training recommendations and tools.

Table 8-1. Cultural education recommendation

	Training Recommendations
Cultural Education	• Comparison of cultural values and social structures (United States compared to those of the operational area). • Local customs and traditions (for example, greetings do's and don'ts). • Geopolitical history (pre-colonial to contemporary and the orientation of each faction or party). • The role of religion in daily life.
	Training Tools
Resources	• Guest speakers native to the country of interest (for example, nongovernmental organization staff, foreign students, recent immigrants, or selected refugees). • Others who have worked in or studied the mission area (for example, special forces personnel, diplomats, and scholars). • Cultural familiarization handbooks.
Format	• Combination of briefings, small group discussions, and question and answer periods. • Handouts to augment—not replace—speakers. • Visual media, specifically slides and videos of the mission area.

8-31. Advisors with previous operational experience have addressed the importance of cultural education. Many civilian agencies have developed intercultural effectiveness programs designed for their personnel working overseas. Often these programs can be revised and tailored for military and paramilitary groups. If available, these programs are an excellent resource for predeployment preparation.

CULTURAL GAP

8-32. Each culture has its own societal rules regarding who a person may speak to, how and when the person may speak, and what topics the person may speak about. Many cultures rely on nonverbal signals to communicate. In such cultures, posture, expression, and actions convey more than spoken or written words. Advisors work to bridge the cultural gap by learning the language as well as nonverbal communication cues. To bridge the gap, advisors develop a sense of cultural awareness, recognize differences, learn to adapt, develop tolerance, and maintain personal contact.

Develop a Sense of Cultural Awareness

8-33. Advisors must recognize they are a product of their own culture. They must learn as much as possible about the culture of the people with whom they need to communicate. When communicating with people across cultures, advisors abandon any sentiments of ethnocentrism—the tendency of individuals to judge all other groups according to their own group's standards, behaviors, and customs. Such notions lead an individual to see other groups as inferior by comparison.

Recognize Differences

8-34. Each culture has its own way of accomplishing required daily tasks. Advisors must understand that each particular society may approach things differently; it does not mean that they are inefficient or less valid. Differences should not be seen as negative. Respect for counterparts must be maintained at all times.

Learn to Adapt

8-35. Advisors must be flexible and ready to adapt and adjust their behavior; however, they must be careful not to overdo their adjustment. Individuals who are overly flexible are often perceived as being insincere. The successful advisor must strive to act in a way that is appropriate to the target culture. Above all, advisors should be themselves and show sincerity.

Develop Tolerance

8-36. Advisors must develop a tolerance for deviations from accepted norms. Events or activities that may seem extraordinary to newcomers may be common practice in the culture. Advisors must be aware that members of the foreign culture may be astounded by that which is commonplace in the United States. Careful observation should be made before judgments are rendered about seemingly peculiar behavior.

Maintain Personal Contact

8-37. Personal contact is the most effective way to bridge organizational barriers. Interorganizational stumbling blocks are very real and the prejudices that arise from them are exacerbated by misunderstanding and ignorance. Often it is too easy to attribute negative attitudes and hostile motives to faceless groups. The advisor must employ superior interpersonal skills and work directly and closely with individual members of other organizations. This type of contact can effectively reinforce commonality and diminish the impact of disparity. These personal relationships are the key to effective interorganizational relationships.

Culture Shock and Adaptation

8-38. The term culture shock describes the anxiety experienced by persons in a completely new environment. It is not the result of a single event. Rather, the condition develops slowly from a series of minor events or conditions. Culture shock expresses the feelings of not having a set direction, not knowing what to do or how to do things in a new environment, and not knowing what is appropriate or inappropriate.

8-39. Culture shock occurs because the mind and body have to go through a period of psychological and physiological adjustment when individuals move from a familiar environment to an unfamiliar one. The cues received by all of the senses suddenly change. During the day the foreigner is bombarded with unfamiliar sights, sounds, smells, tastes, languages, gestures, rules, requirements, interactions, demands, systems, and expectations.

8-40. Some differences that advisors experience between their lives at their home station and their lives when deployed to a foreign location are obvious. These differences include language, climate, religion, food, educational system, and the absence of family and friends. Other differences may not be as obvious. These differences include how people make decisions, spend their leisure time, resolve conflicts, express their emotions, and use their hands, faces, and bodies to express meaning.

Stages of Shock

8-41. Five distinct stages occur in the culture shock process. These include enthusiasm, withdrawal, reemergence, achievement and reentry. Figure 8-1 depicts the stages of shock.

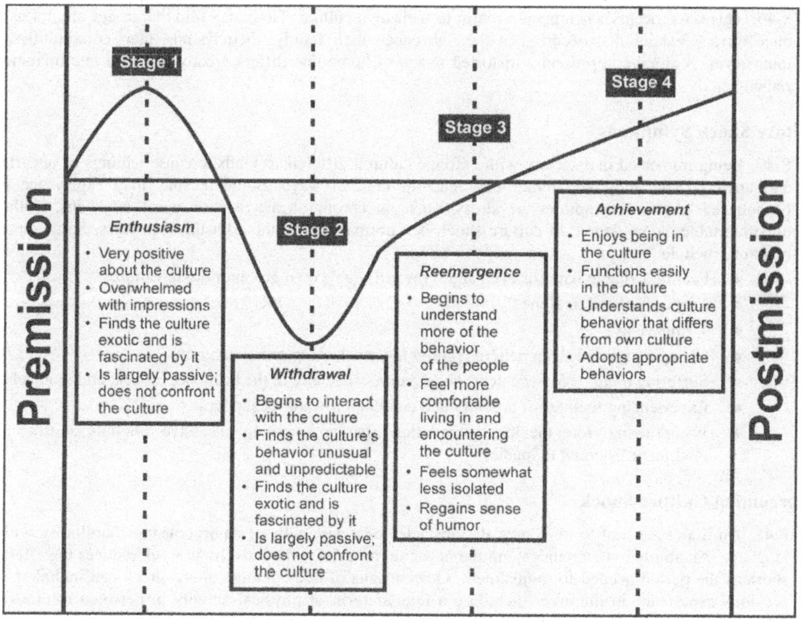

Figure 8-1. Stages of culture shock

Stage 1 – Enthusiasm

8-42. In this stage, a person may feel self-confident and pleasantly challenged. The person may be delighted by all of the new things encountered.

Stage 2 – Withdrawal

8-43. In this stage, a person begins to encounter difficulties and minor crises in daily life. It may be difficult to make oneself understood and foreign counterparts may prove more difficult to work with than anticipated. In this stage, one may have feelings of discontent, impatience, anger, sadness, and even incompetence.

Stage 3 – Reemergence

8-44. This stage is characterized by gaining some understanding of the new culture. A person may not feel as isolated, and a feeling of direction emerges. The individual is more familiar with the environment and is better able to adjust to the circumstances.

Stage 4 – Achievement

8-45. In this stage, the person realizes that the new culture has both good and bad things to offer. This stage is one of integration and the person is increasingly able to function in the new setting. A sense of accomplishment, a reduction of routine annoyances, and a more solid sense of belonging accompany this integration.

Stage 5 – Reentry

8-46. This stage occurs when people return to their own culture. They may find that things are not as they once were. Changes that occurred in their absence—their family, their friends, their communities, and themselves—combine to present a distorted image of home that differs greatly from the one imagined or remembered.

Culture Shock Symptoms

8-47. Being immersed in a society with extreme cultural differences tends to cause feelings of uncertainty and anxiety. The body and mind may react in unusual ways. Some persons may experience more pronounced physical symptoms of stress, such as chronic headaches or upset stomachs. Although uncomfortable, some degree of culture shock is a normal part of the adjustment process. Some common reactions include:

- Feeling irritable with (or even anger toward) one's own group or organization.
- Feeling isolated or alone.
- Tiring easily.
- Changing normal sleep patterns (either too much sleep or not enough).
- Suffering minor (but persistent) body pains, especially in the head, neck, back, and stomach.
- Experiencing feelings of hostility and contempt toward local people.
- Withdrawing from the local population (that is, spending excessive amounts of time alone reading or listening to music).

Overcoming Culture Shock

8-48. An important tool to overcome the obstacles of a new cultural environment is familiarity with the language. An ability to communicate in the new culture, even at the most basic level, reduces the effect and shortens the period needed for adjustment. Other means of overcoming culture shock can include having previous experience in the area (including a regular form of physical activity in personal routines) and acknowledging even slight progress in adjusting to the new culture.

8-49. TC 31-73 discusses culture shock in depth. It identifies the symptoms, cues, and treatment.

CROSS-CULTURAL COMMUNICATIONS

8-50. Communication is the transfer of messages from one person to another. These messages may be passed along verbally, in writing, or by signals (gestures, body language). The sender encodes the message and the receiver decodes it. The type and style of encoding used is based upon the sender's history, beliefs, values, opinions, and belief systems. Communication is determined by culture; as a result it can be either high context or low context. High context communication places an emphasis on clearly stating issues and problems and working towards resolving them in a straightforward manner. Low context communication places a premium on building relationships; as such, problem sets tend to be worked in a less formal matter in societies that utilize this type of communication. Quickly and effectively identifying whether a society uses high or low context communication is a key to effectively conveying material between the advisor and the host-nation population.

8-51. The receiver decodes messages based upon lifestyle, group membership, status and role, worldview, language, and social norms. Communication is a two-way process in which the encoding and decoding methods can affect both sending and receiving. Effective communication occurs when the message is perceived and responded to in the manner the sender intended. Ineffective communication can occur from poorly chosen words, flawed timing, a confused mixture of verbal and nonverbal signals, poor listening skills, and a failure to take culture into account. Communications can be broken into both levels and factors.

Levels of Communications

8-52. The cross-cultural communications capabilities required by an effective advisor to FSF can be described in three levels. At the lowest level is awareness followed by knowledge. The highest level is reached when these two are combined with well-trained and refined skills. TC 31-73 discusses communications in detail.

Awareness

8-53. Awareness is the basic level of cross-cultural capability. Awareness of cultural differences and their impact is the first prerequisite for successful work with a counterpart. Simply being sensitive to the fact that differences exist and carefully observing actions and reactions can assist the advisor in adjusting behavior and modifying actions to achieve greater influence with the counterpart. Awareness is not region specific and can be instilled in the advisor with relatively little training.

Knowledge

8-54. Knowledge of the details and nuances of a specific culture is the next level of cross-cultural capability. Advisors attain this level through academic study and immersion. Such knowledge is inherently area-specific and rarely transfers from one target area (or culture) to another. Developing in-depth regional knowledge necessary for effective cross-cultural communications requires extensive and time-consuming training. Appropriate personnel assignment policies must support this training.

Skills

8-55. Skills fundamental to effective cross-cultural communications, when combined with awareness and knowledge, form the highest level of cross-cultural capability. These skills include professional competence, language, nonverbal communication, negotiation, interpersonal skills, observation, problem solving, leadership, instruction, and fitness. Although some individuals show greater natural talent for these skills than others, all advisors require continual training in each to achieve and retain their full potential as advisors to foreign counterparts. These skills are discussed in more detail in TC 31-73.

Factors Affecting Communication Across Cultures

8-56. There are four aspects that can impact communication across cultures. These factors are level of direct communication, perception of time, perception of the individual versus the group, and show of emotion.

Level of Direct Communication

8-57. Cultures fall along a spectrum of formality, with some cultures being direct and explicit in their communication and others less direct and more ambiguous. Cultures that rely on direct communication place more emphasis on verbal communications that are precise and explicit. Context and nonverbal cues are secondary to verbal communication. Cultures that rely on indirect communication rely less on verbal communication and more on context and nonverbal cues. Their verbal communication can seem vague and noncommittal to those from a culture that is more direct.

Perception of Time

8-58. Cultures can differ greatly in their perception of time. This difference divides into two cultural concepts of time: monochronic-time and polychronic-time. In monochronic-time cultures, members emphasize schedules, precise reckoning of time, and promptness. In such cultures, the schedule takes precedence over the interpersonal relation. In polychronic-time cultures, time is viewed as fluid. Members of polychronic-time societies do not observe strict schedules—agendas are subordinate to interpersonal relations.

Perception of the Individual Versus the Group

8-59. Cultures may be individualistic or collectivist in their orientation. An individualist culture is one in which the ties between individuals are loose—where people are expected to take care of themselves and their immediate families. In a collectivist culture, people are raised from birth into strong, cohesive groups. These groups offer a lifetime of protection in exchange for unquestioning loyalty. Advisors must understand from which culture their counterparts come to identify motivations accurately and communicate effectively.

Show of Emotion

8-60. Each culture has its own system of expressing emotion. Some cultures tend to express their emotions and show their feelings openly. Other cultures tend to repress those feelings more. Rather than showing their feelings openly, members of these cultures keep their emotions controlled and subdued. Those from the more expressive culture may view people from the repressive culture as cold or unfeeling. Similarly, those from the more repressive culture may view their more expressive colleagues as eccentric. To communicate effectively, advisors must be able to understand their foreign counterparts' emotional communication and display their own emotions carefully.

Chapter 9

Working with Counterparts

Rapport is a relationship marked by cooperation, conformity, harmony, or affinity. When people discuss good rapport, they describe a relationship founded on mutual trust, understanding, and respect. Relationships characterized by personal dislike, animosity, and other forms of friction often lack rapport. For the advisor, rapport describes the degree of effectiveness.

ESTABLISHMENT OF RAPPORT

9-1. Since advisors are in a unique military position, they establish rapport with their foreign counterparts. That position is one in which the advisor has no positional authority over the actions of their foreign counterparts. This lack of authority means that the doctrinal view of leadership is modified to emphasize interpersonal relationships and deemphasize authoritarian roles. Advisors instead use their people skills to influence the actions and decisions of their counterparts. Advisors use their people skills to building rapport. They have a genuine interest in other people, smile, remember and use people's names, encourage others to talk about themselves, listen to others, discuss what the other person is interested in, and make the other person feel important.

9-2. Advisors obtain certain knowledge before establishing effective rapport. First, they study FM 6-22 to gain basic leadership knowledge for understanding human nature and motivation. Advisors then incorporate information specific to the culture and society of their potential counterparts. This information may take the form of thorough area studies, operational area studies, and other research materials. Knowledge includes the components and techniques of rapport.

COMPONENTS OF RAPPORT

9-3. Three components compose rapport: understanding, respect, and trust. All the components of rapport are two-way streets; to have counterparts share about their culture, language, and experiences, advisors have to be willing and able to share also. Building this rapport may require the advisor to establish a personal level of understanding with their counterparts. Understanding is the first component of rapport. It begins before deployment and may include foreign cultural studies, language training, and equipment and doctrine familiarization. Once in country, advisors continue to broaden their understanding by observing and asking questions.

9-4. Respect is the next component of rapport. It is a reciprocal commodity. The foreign counterpart should grow to respect advisors, what they know, and how they perform. Advisors should look for those things that they respect in their counterparts. This may prove challenging. The counterpart may not fit the mold of U.S. or other coalition officers. Lacking formal training or education, the foreign counterpart may even be illiterate. The uniform standards of the host nation often may differ from the advisors' standards. Start with the basics and expand over time. Additional factors may affect the counterpart's willingness and motivation. For example, by accepting the duties, the counterparts (and their family) may be in mortal danger. The counterpart may have proven to be a fearless warrior, even without formal training. In short, traits deserving respect will exist even before the relationship matures. Mutual respect grows through shared experiences and shared dangers. Advisors should live, eat, and if authorized, fight with their counterparts. By sharing their hardships and dangers, advisors build respect.

9-5. Trust is the final critical component for building rapport. Trust grows gradually from understanding and respect. Building trust needs to begin on day one, but it will not mature until later in the relationship.

Advisors should begin by showing they are reliable and should do everything they say they will do. They should be in the right place at the right time. Two things can enhance this growth. First, the advisor should start out with confidence-building missions. Early success builds self-confidence and trust. Second, advisors should not promise any support they might not be able to deliver.

TECHNIQUES OF RAPPORT

9-6. Foreign security forces (FSF) respect advisors who actually show them the right way to do a task rather than just talk. Experience has shown that foreign units will not hesitate to go into danger if their advisors are right there with them. To further contribute to building rapport, advisors should maintain enthusiasm and a positive attitude; learn the language or how to use an interpreter; respect a counterpart's rank, age, status, and experience; develop negotiation skills; and improve interpersonal skills.

9-7. Other rapport-building techniques can include being confident, competent, and capable. Advisors never expect or demand their counterparts to do something that they are unable or unwilling to do. They demonstrate everything. Advisors learn their counterparts' names and spend time with them at meals and during holidays; they prepare mentally to interact with counterparts at all times. Advisors avoid creating an American-only enclave; they avoid giving the impression of favoring any one group. They frequently assess their counterparts' perceptions. Advisors recognize threats to discipline and enforce the chain of command.

ACCOMPLISHMENT OF RAPPORT

9-8. Advisors measure rapport by how well they can motivate their counterparts to act. The basic techniques of motivation (in the absence of authority) are advising, setting the example, seeking compromise, and coercing. TC 31-73 discusses rapport in detail.

ADVISING

9-9. Advising the counterpart to select a particular course of action is only effective if the counterpart trusts the advisor is professionally competent. If the counterpart does not perceive the proposed solution as realistic, the advisor's competence is questioned. Advisors carefully explain to their counterparts that recommended courses of action are realistic and will be effective.

SETTING THE EXAMPLE

9-10. Setting the example for the counterpart must be an ongoing effort for the advisor to avoid the appearance of a "do as I say, not as I do" attitude. In setting the example, advisors explain that what they are doing is the most effective action for the situation. This is particularly true when the counterpart really does not understand the behavior (or purpose).

SEEKING COMPROMISE

9-11. When seeking compromise with the foreign counterpart, advisors may create a situation in which the counterpart has a personal interest in successful execution. In some cultures, seeking a compromise may allow the counterpart to save face. Furthermore, in certain situations the counterpart—because of practical experience—may have a better solution to the problem at hand.

9-12. When seeking a compromise in certain cultures, the perceived competence of the advisors may suffer. This may be mitigated somewhat by approaching the compromise as two professionals (the advisor and the counterpart) reaching a mutual conclusion. To reach an effective compromise, advisors may have to conduct negotiations.

Note: Two areas of concern must never be compromised for the sake of maintaining rapport—operations security and human rights.

COERCING

9-13. Coercion is the least desirable method of motivation because it can cause irreparable damage to the relationship. Advisors should use coercion only in extreme circumstances. Advisors must avoid unintentionally forcing their counterparts into action. Instead, they establish and maintain good rapport by conveying they are sincerely interested in their counterparts, their nation, and its cause. Advisors will not belittle their counterparts' efforts. Advisors must demonstrate that they have come to help because they believe the counterparts' goals are just, fair, and deserving of success.

POLICIES AND AGREEMENTS

9-14. Advisors must remember their primary aim is to forward U.S. policy. Therefore, they establish relationships with foreign personnel, other U.S. forces, and other government agencies. Advisors promote U.S. policy through agreements, command relationships, and operations. TC 31-73 discusses policy in greater detail.

UNITED STATES AND HOST-NATION POLICIES

9-15. Advisors not only support U.S. national policy in the areas where they operate, they forward the policies of the host-nation government (unless otherwise directed by higher authority). Advisors maintain communications with higher headquarters to ensure activities align with U.S. objectives.

9-16. Host-nation policies may dictate practices and procedures that appear inefficient or uneconomical. Advisors should avoid criticizing or condemning such practices and procedures until they thoroughly understand them. Advisors should make private recommendations critical or contrary to host-nation policy.

AGREEMENTS

9-17. The status of advisors is normally specified by a status-of-forces agreement or other arrangement between the U.S. and host-nation governments. These agreements vary widely; they may offer little political protection or may provide for full diplomatic immunity. If no such agreement exists, advisors are subject to local laws and customs. Even if a status-of-forces agreement or other agreement exists that provides immunity, advisors are still expected to observe local laws.

COMMAND RELATIONSHIPS

9-18. Advisors must use the U.S. chain of command to obtain and disseminate guidance and assistance. They carefully distinguish between the U.S. and foreign security forces' chain of command. In particular, advisors prevent their counterparts from attempting to control subordinates through the U.S. chain of command. Advisors provide recommendations—not orders—to their counterparts. Only the counterpart should issue orders to subordinates.

9-19. Nongovernmental organizations and other government agencies may be co-located in the operational area. Advisors integrate the efforts of all organizations into their mission planning and should seek opportunities for integrated efforts with their counterparts. This type of unified action at the tactical level can be achieved through the planning process to integrate nonlethal actions. Holistically, nonlethal actions include development and diplomatic efforts.

CONDUCT OF OPERATIONS

9-20. Advisors can expect to participate in stability operations. These operations use military and paramilitary forces, as part of the interdepartmental team, to support internal defense and development operations. U.S. assistance may include advice on military organization, training, operations, doctrine, and materiel. In addition, the United States may provide and control U.S. logistics and sustainment for FSF. The objective of this assistance is to increase the capability of foreign organizations to perform their missions and operate efficiently in the given operational environment.

LEVELS OF ADVISORS

9-21. Organizations and individuals possessing greater skills and resources assist by imparting their knowledge through assistance efforts. The success of assistance depends largely on effective interaction between advisors and their foreign counterparts. Advisors may operate in the national, operational, and tactical levels. The national level is considered the largest national subdivision. This may be a nation's combined military headquarters, Service components, national-level government officials, or a combination. Operational-level advisors work at a level below the national-level advisor. This may be with military corps or division officers, with regional government officials, or a combination. Depending upon the size of the nation and its military, operational-level advisors may be assigned as low as brigades. Tactical-level advisors normally advise military units at or below the brigade level and provide counsel and assistance across the spectrum of conflict. They advise and assist counterparts in developing unit combat effectiveness and serve as liaisons between FSF and U.S. combat, support, sustainment and protection forces. Chapter 3 of TC 31-73 discusses these levels in detail.

CONSIDERATIONS OF RAPPORT

9-22. When building and maintaining rapport, advisors consider intelligence, human rights, role shock, ambiguity, personal relationships, bureaucracy, responsiveness, and diversity. These considerations are discussed in greater detail in TC 31-73.

INTELLIGENCE

9-23. To promote national policy and attain overall objectives, advisors may need to plan and conduct intelligence training. The advisor should assist the counterpart in developing a local intelligence collection program, training intelligence personnel in their respective specialties, and utilizing trained intelligence personnel properly.

9-24. The advisor should be familiar with area studies and area assessments. Pertinent documents should be compared to detect trends or changes. The advisor also should evaluate the foreign intelligence staff, its standing operating procedures, its chain of command, intelligence projects, and reference material available from other intelligence agencies. Advisors should prepare and maintain a list of essential elements of information.

HUMAN RIGHTS

9-25. Advisors stress the consequences of mistreating suspects, prisoners, or other persons taken into custody. These persons must be treated per Article 3 of the Third Geneva Convention. Article 3 requires care for the sick and wounded, protection of prisoners and detainees of all types from abuse or other harm. Murder, mutilation, and torture are expressly forbidden, as is humiliating or degrading treatment. Advisors must never be active participants in the conduct of such punishment. Furthermore, they must explain to their counterparts that they are obliged to report any atrocities of which they have knowledge.

ROLE SHOCK

9-26. Role shock results from the discrepancy between the roles individuals expect to play and the roles they actually do play. Role shock may also be a product of the tension created by individuals trying to do jobs themselves versus advising others how to do the jobs. Role shock tends to increase until about the anticipated midpoint of the tour and seldom disappears completely until redeployment. Symptoms of role shock resemble those produced by other stressful life events.

9-27. Many advisors become entangled in complexities of bureaucracy—both foreign and American—and become increasingly frustrated when seemingly familiar things fail to respond in expected ways. This continuing frustration can be a key contributor to role shock.

9-28. Although most advisors undergo some degree of role shock, over time they learn to take the personal, living, and social conditions in stride. More challenging are the problems that arise in connection with their jobs—their professional roles, relationships, personal achievement, self-development, self-determination, and self-image. Often advisors find their duties and activities during deployments differ somewhat from those expected. The actual duties and responsibilities are greater in scope, involve technical work outside of their assigned specialties, and require honed administrative (rather than military) skills.

AMBIGUITY

9-29. Most advisory assignments are ambiguous. What an advisor is expected to achieve is not always made clear. Furthermore, assignments may conflict with what they have trained to do, what others expect of them, or what they want or expect to do.

9-30. U.S. and foreign security forces may spend months—even years—negotiating a project. Still, the overall objectives or definition of the mission to be accomplished might be expressed in general terms. Various interpretations of what is to be accomplished arise. Even more views surface regarding the most appropriate means of reaching those goals.

RELATIONSHIPS

9-31. Ideally, foreign counterparts provide a working context for advisors and help them to understand and adjust to the new culture. Historically, those with good working relationships with their counterparts make better adjustments than those who have no counterparts or bad relations with their counterpart. Good working relationships may be difficult to establish and maintain. Most advisors approach their assignments as teachers and mentors. Foreign counterparts may resist such a relationship, particularly if they were expecting someone to take over and do all or part of the job. In addition, many counterparts want their advisors to share in at least some of the blame if something goes wrong.

9-32. The United States has little control over the availability or capability of counterparts. The assigned individual may be technically incompetent, lack professional commitment, have other interests, or be unavailable when needed. In some cases, numerous counterparts are assigned and rotate in their workings with the advisor. A single counterpart may not be assigned until the advisor has been on the scene for some time. In some scenarios, a specific counterpart is never assigned.

BUREAUCRACY

9-33. Although some Americans might disagree, the United States maintains a system that is—by world standards—relatively permissive and nonbureaucratic. As such, advisors may become aggravated when confronted with or obstructed by host-nation bureaucracy, red tape, administrative centralization, and a cumbersome decisionmaking process. Cultures that emphasize consensus in the decisionmaking process often prove quite frustrating for advisors. To be successful, advisors strive to understand the bureaucracy in which they operate.

9-34. Advisors should recognize the value of being or knowing the middleman. This role is sometimes one of the key contributions an outsider can make. Middlemen can move in, around, and out of organizations somewhat independently of the local hierarchy, protocol, and customs. They can link persons and offices that might otherwise have difficulty communicating. Many advisors dislike this role and refuse to practice it; however, the middleman can be a significant catalyst in any advisory operation.

RESPONSIVENESS

9-35. As frustrating as foreign bureaucracies may be, advisors often get even more irritated with the responsiveness of U.S. forces. Advisors frequently fail to account for distances, lead times, diplomatic issues, and other challenges when conducting security force assistance. Moreover, few Soldiers in the unit conducting security force assistance have had substantive experience in working in an interagency environment. These issues are further complicated by the usual tensions between headquarters and field staffs and between administrators and technical specialists. Expectations, values, and styles of behavior vary from group to group.

9-36. Deploying personnel may not be familiar with the tasks of development and the vast range of problems typical of underdeveloped areas. They come in eager to contribute their knowledge and skill, and anxious to make maximum progress in the limited span of their assignment. In so doing, they may fail to recognize that the factors that frustrate them—such as the lack of professional skills and facilities—are actually their reason for being there.

DIVERSITY

9-37. Organizations attract individuals who differ in abilities, knowledge, skills, work ethics, social behaviors, values, and attitudes. Advisors must be able to recognize personality types and develop techniques to cooperate effectively with each type.

RECOMMENDED PRACTICES IN BUILDING RAPPORT

9-38. Advisors may adopt many practices to help work with counterparts. Practices that help advisors build rapport are discussed in paragraphs 9-39 through 9-51.

ENSURE THAT THE ADVISOR PRESENCE IS UNDERSTOOD

9-39. Advisors entering an operational area work under the sponsorship of a military commander, mayor, village head, or some other recognized local leader. Prior to this, the advisor's arrival may need to be coordinated through the district, sector, or other appropriate local office. Clearances from distant national, state, province, or sector governments cannot compensate for a clear, local explanation of why an American advisor is in the area. This particularly applies to small, isolated communities where it is unusual for a stranger to appear for even an hour without being acknowledged and accepted by local leaders. Without explanations from locally respected persons, civilians often arrive at their own explanations, often to the detriment of the advisor effort.

FIND A BASIS FOR COMMON INTEREST WITH THE LOCAL PEOPLE

9-40. If advisors show appreciation for civilians as individuals, cultural gaps and language barriers can be overcome and common ground can usually be found. Advisors should listen and show interest when the locals speak. Initial conversations usually center on universal matters, such as food, shelter, clothing, health, and education. Over time, discussions may naturally move to the matter the advisor wants them to consider. Advisors are better received if they know something about earlier contributions in such matters as agriculture, folk art, religion, and architecture. Naturally, advisors prove more effective and appreciated if they can speak the language.

TRY TO UNDERSTAND WHY THINGS ARE DONE THE WAY THEY ARE

9-41. Although some local practices may seem strange, they generally have good reasons behind them. Advisors can discover these reasons with careful observation and a creative imagination. Food habits, family traditions, folk cures, and festive celebrations almost always have a great deal of human experience at their root. The advisor also needs to be aware that many villages contain rival subgroups and factions. This tension needs to be accounted for when working with the local people.

START WITH WHERE THE PEOPLE ARE AND WHAT THEY WANT

9-42. Advisors find out what the local people really want and work with them to achieve this aim. The need felt by the local people often is the best starting point. Once advisors address this aim, people become more appreciative and cooperative—they begin to raise their expectations and become interested in working for other improvements. To address the initial desires of the people, the advisor may need to call in personnel with the specialized skills needed for the particular project. Although this can result in a delay, it will help the advisor to achieve better cooperation.

Work Within the Local, Cultural Framework

9-43. Advisors understand basic cultural matters such as ethnic backgrounds of the people, family relationships, leadership patterns, and value systems. They also need some knowledge of local services such as health, education, communications, and transportation.

Help People Believe They Can Improve Their Situation

9-44. Most traditional peoples of South America, Asia, and Africa live in a largely static environment. Through their experience with change, they are more fearful of losing status than they are hopeful of bettering their condition. Therefore, changes suggested by advisors are often viewed with fear. Concrete local projects that yield easily observed benefits help convince villagers that they can improve their situation and make them more willing to cooperate in other projects.

Be Content with Small Beginnings

9-45. Change tends to come slowly in areas where there have been few in recent times. Advisors remember that knowledge—technical or otherwise—is cumulative. Once a small beginning has been made, greater activity and additional changes soon follow. It is easier to achieve momentum than to maintain it. Regardless, advisors start within as promising a framework as possible and with the support needed to sustain the momentum achieved.

Utilize Local Organizations and Recognize Their Leaders

9-46. The surest way for an activity to be continued after the advisor leaves is for it to have been launched and carried forward within the local organizational and leadership framework. People respond best when their local organizations are recognized as important and useful. Advisors consult and encourage recognized local leaders to contribute as they can. A well-conceived technical activity reflects credit on the local leaders associated with it. Advisors give attention not only to officials and family heads of local groups, but also to the quiet, behind-the-scenes leaders.

Help the Government Get Organized To Serve The People

9-47. For advisors to be most effective, they must understand the organization of the local government and how their activities fit into the overall scheme. Usually a set of agreements exists between various local agencies and the national government (usually through some sort of interministerial council). This agreement provides for a coordinated effort in servicing the varied needs of civilians. The advisor may need to work with appropriate agencies to assist in getting such agreements made. If such agreements already exist, advisors should be careful to recognize and strengthen them.

Train and Use Subprofessional, Multipurpose Local Workers

9-48. Selected young people in the villages can be trained and used as subprofessional, multipurpose village workers. Otherwise, advisors' influence is restricted to where they are standing and the immediate vicinity. Furthermore, advisors may spend so much time establishing and maintaining enough rapport with the villagers that they may become incapable of rendering any real service at all. Often the gap between civilians and the advisor is formidable because of their educational and cultural differences. Often the advisor works with villagers who are poor, illiterate, and devoid of outside contacts. Volunteer or paid local workers have proven quite helpful as they serve as a liaison between villagers and advisors.

Expect Slow Progress

9-49. As civilians begin to see successes from their joint efforts and begin to have new hope, they naturally want a larger hand in their own matters. The advisor may sometimes feel they want to assume more responsibility than they can carry. These evidences of growing pains should be appreciated, for they are a necessary part of becoming able to assume responsibility. The advisor needs to adjust to the growing desires of the people to help themselves.

Transfer Controls Constructively

9-50. Advisors need to help civilians see how they can build the new—what they want—upon the foundations of the old—what they already have. From the beginning of a project, the advisor's team needs to talk with local leaders. Together they envision how to train local personnel, gather financial support, and transfer responsibilities before transferring control of activities. If operating responsibility is transferred too early, some breakage—usually of material things—will likely occur. If the team keeps control too long, the civilians who wanted to take over may grow disillusioned with them for not relinquishing control when they thought it should have been. Working with local leaders, the team determines when to phase out each technical activity.

Expect Little Gratitude from Those Helped

9-51. People who benefit from assistance sometimes feel defensive. In accepting assistance they are, in a sense, admitting their own insufficiency. The self-esteem of a person, community, or a nation as a whole is delicate. The advisor's team should not, therefore, expect thanks. Rather, the team should approach the people in a spirit of fraternity and humility, taking satisfaction in their forward progress. The team should do its job the best it can and accept work well done as is its own reward.

DYNAMICS OF ADVISING FOREIGN SECURITY FORCES

9-52. Advisors work in a complex environment that is heavily influenced by the unified effort of multiple organizations, relationships, compartmentalization, and organizational cultures. Paragraphs 9-53 through 9-57 discusses the probable dynamics that an advisor must take into account.

Advisors and Interdependent Operations

9-53. Advisors commonly operate in complex environments with interdependent operations. Interdependent operations require near-seamless integration of agencies representing the instruments of national power—diplomatic, informational, military, and economic. This same degree of integration must be extended to include external organizations and those agencies representing the instruments of power in partner states. The negotiation and cross-cultural skills that enhance their performance as advisors are also well suited to the complex interrelationships that characterize such operations.

Organizational Relationships

9-54. Often organizational relationships can often be misleading and must be clarified. The actual interrelationships among and within organizations seldom follow a line-and-block diagram. Instead, they are heavily influenced by circumstances, personalities, perceptions, and resources. All relationships and lines of authority are subject to negotiation. Advisors understand that clarifying the roles, functions, and responsibilities is a continuous process and that each statement or action sets a precedent for future interrelationships.

9-55. The advisor must become hypersensitive to the most minor organizational disparities. The apparent similarities—the common language, sociological backgrounds, and citizenship—can be disarming and lead the advisor to overlook very real differences. If ignored, these differences can quickly impede synchronization.

Compartmentalization

9-56. Advisors balance security and synchronization. Overcompartmentalizing must be avoided. Often distrust can be prevented simply by not being overly secretive. Although advisors must conduct a careful, continual risk assessment to avoid compromise, the sharing of critical information is necessary for harmonious operations.

Organizational Culture

9-57. Advisors should approach external organizations as they would foreign counterparts—as unique institutions with distinctive cultures. Organizational differences, biases, and approaches must be researched and—whenever possible—accommodated. Each Service in the Department of Defense has unique perspectives that lead to significant differences in objectives and approaches to various problems. Profound differences in perspective may even be found between the different branches of the Army. If not recognized and accommodated, these characteristics may create distrust, disharmony, and disruption during critical operations.

Chapter 10

Cross-Cultural Influencing and Negotiating

Leveraging foreign counterpart cooperation can prove a difficult and critical task. Although coercion may prompt surrogate forces to perform a specific action, it can never lead to established conduct and legitimacy. Lasting advisors can achieve long-term success only by patiently applying their influence.

Advisors analyze their counterparts and determine their motivations and needs. Basic needs, culture, and individual personalities influence people. Advisors determine what these three things are so the appropriate influence techniques can be applied. Then advisors negotiate. Negotiation is fundamental to problem solving using the study, development, and honing of negotiation skills critical to the advisor.

INFLUENCING

10-1. Advisors need to gain influence with their counterparts, leaders, civilians, coalition members, and others beyond their chain of command. Advisors are self-aware, nonjudgmental, and capable. It is not necessary to accept others' beliefs or values, but advisors need to understand others' ways of thinking and operating. Knowing their own culture and understanding why they believe and value what they do helps advisors find a common cause. A common cause unites a diverse set of decisionmakers with whom advisors interact and try to influence.

10-2. Once advisors gain a minimum amount of influence, they can make things happen through others. They can extend their influence beyond their chain of command. Influence is simply getting others to do what you want.

10-3. Advisors identify the types and principles of influencing. They determine the best techniques with which to influence and review vital considerations. Using measures of effectiveness, they check that their influences have the desired results.

TYPES

10-4. Commitment, compliance, and resistance clarify how advisors understand the impact of their attempts to influence their counterparts. These three types of influencing are determined by how committed the foreign security forces (FSF) are to the task.

Commitment

10-5. Commitment is a willing dedication to a requirement or a cause because one wants to. Commitment-focused influence generally produces longer lasting and broader effects. Whereas compliance only changes immediate behavior, commitment reaches deeper—changing attitudes, beliefs, and behavior. For example, when advisors build desire and responsibility for a task among FSF, they often demonstrate more initiative, personal involvement, and creativity.

Compliance

10-6. Compliance is conforming to a requirement or demand because they feel like they have to avoid punishment or to gain a desired reward. Compliance-focused influence is based primarily on the leader's authority. Giving a direct order to a follower is one approach to obtain compliance during a task.

Compliance is appropriate for short-term, immediate requirements and for situations where little risk can be tolerated. Compliance techniques are also appropriate for leaders to use with others who are relatively unfamiliar with their tasks or unwilling or unable to commit fully to the request. If something needs to be done with little time for delay, and there is not a great need for the target of influence to understand why the request is made, then compliance is an acceptable approach. Compliance-focused influence is not particularly effective when a leader's greatest aim is to create initiative and high esteem with a counterpart.

Resistance

10-7. Resistance is a reaction against change. Resistance generally takes one of four forms—reactance, distrust, scrutiny, and inertia. It is manifested behaviorally (I won't do it), affectively (I don't like it), and cognitively (I don't believe it). Recognizing and lowering potential sources of resistance are key components to gaining compliance and commitment.

MOTIVATIONS

10-8. Different motivations influence behavior. One type of motivation is the desire for something. Another is the fear of something undesirable. As an agent of influence, advisors determine how to increase approach motivations and how to decrease avoidance motivations. For instance, recruiting for a special mission can use special badges as incentive (approach). Advisors can assure Soldiers that the mission is relatively safe by providing extra training and better equipment (lowering risks or avoidance).

PRINCIPLES

10-9. Six influencing principles are relevant to virtually any culture and any group. Understanding and applying these principles can increase an advisor's effectiveness in persuading a group or individual:

- Principle of scarcity.
- Principle of authority.
- Principle of social proof.
- Principle of liking.
- Principle of reciprocity.
- Principle of consistency.

10-10. With the principle of scarcity, people value more of what they can have less. They typically associate greater value with things that are rare, dwindling in availability, or difficult to acquire.

10-11. In the principle of authority, people are more easily persuaded by individuals perceived to be legitimate authorities or experts. They defer to experts who provide shortcuts to decisions requiring specialized information.

10-12. With the principle of social proof, people often look to the behavior of those around them for direction about what choices to make. This action is heightened when those around them are similar in terms of age, education, social standing, and experience.

10-13. In the principle of liking, people prefer to say yes and to comply with the requests of those they like.

10-14. In the principle of reciprocity, if someone grants favors, invited or uninvited, an overpowering need to repay that favor immediately blooms within the recipient of that favor. This human trait transcends all cultures and races.

10-15. With the principle of consistency, the desire for consistency motivates behavior. The drive to look and be consistent is a highly potent weapon of social influence, often causing people to act in ways clearly contrary to their own best interests. People do not like to appear inconsistent to others. It is human nature that people strive to feel good about themselves, which includes behaving per their important values and beliefs. When the behavior is consistent with whom people are and what they value, they feel good. People align with their clear commitments.

TECHNIQUES

10-16. The art of influencing involves knowing what techniques to use based on the situation. Advisors have to know the limits to which can use their influence; their influence is finite. They need to know when to use their influencing skills to achieve the highest payoffs. Their counterparts will only accept so much influencing if they don't completely trust their advisors. Table 10-1 lists positive and negative techniques delineated in FM 6-22. Table 10-2 lists resistance-based influence techniques. These are used when the counterpart resists the advisor's influences.

Table 10-1. Influencing and rapport techniques

Technique	*Description*	*Characteristics*	*Outcomes*
Pressure	Explicit demands, persistent reminders, sometimes appropriate in high stakes, urgent situations	Frequent checking Nagging	Gains compliance Frequent use may cause resentment, undermine morale, and harm relationship
Legitimate Request	Source of authority is used	Source credibility is important	Gains compliance
Exchange	Quid pro quo Reward offered for task completion; punishment threatened for incomplete task	Offer of reward or punishment not tied directly to getting job done, just something that counterpart wants or doesn't want	Gains compliance
Personal Appeal	Request based on friendship or loyalty	Agent of influence highlights talents of counterpart that make target the right person for the job (this is why you are the right person for the job, will you do it for me?)	Gains commitment Builds relationship
Collaboration	Works with target of influence by providing resources of some type, or by removing barriers to task completion	Offer of assistance is directly tied to request	Gains commitment Builds relationship
Rational Persuasion	Using facts, evidence, or logic in a persuasive message	Relies on agent being viewed as expert	Gains commitment Builds relationship
Apprising	Explaining the consequences of a potential action or decision	Decision is good for unit, career, or mission accomplishment	Gains commitment Builds relationship
Inspiration	Arousing strong emotion to build conviction or commitment	Appealing to values of the counterpart	Gains commitment Builds relationship
Participation	Requests for counterpart to be a part of planning, active participation in development of solution	Promotes commitment Possibly more feasible and culturally appropriate course of action	Gains commitment Builds relationship
Relationship Building	Build positive rapport and trust	Provide the foundation for using influence or inspiring	Gains commitment

Table 10-2. Resistance-based techniques

Technique	Description	Characteristics	Outcomes
Sidestep Resistance	Agent defines own role as consultant or advisor, but not commander Pushes the consequence of choice into future Depersonalize task to gain commitment to general idea Focus on end state instead of expended effort	Does not sell, just advises to prevent immediate resistance Puts less focus on immediate impact, more focus on long-term effect Stating "leaders should" instead of "you should" Stress mission accomplishment	Gains compliance Potentially gains commitment if paired with other influence techniques like rational persuasion
Address Resistance Directly	Address reluctance by lowering costs associated with tasks	Provide some of own resources through collaboration or counterargue concerns	Builds commitment
Address Resistance Indirectly	Build confidence, esteem, and self-efficacy for the task	Acknowledge difficulties while showing support for counterpart's abilities	Builds commitment
Distract Resistance	Interfere with the development of counterarguments	Distracting attention	Gains compliance
Disrupt Resistance	Disrupt complacency to bring attention to message	Shake up to motivate	Gains compliance
Consume Resistance	Resistance is energy	Wear them down prior to opportunities to resist. Persist with requests until they concede	Gains compliance
Use Resistance to Promote Change	Reverse psychology	Issuing a challenge	Gains compliance; risky when you don't know target of influence really well

10-17. To succeed with any influence technique, the influence attempt should be perceived as authentic and sincere. Positive influence comes from leaders who do what is right for Army forces, the mission, the team, and each individual Soldier. Negative influence—real and perceived—emanates from leaders who focus more on personal gain and lack self-awareness. False perception may trigger unintended side effects such as resentment of the leader and the deterioration of unit cohesion. With each and every influence attempt, Army leaders should ensure that the influence objective aligns with Army Values, ethics, the Uniform Code of Military Justice, the Warrior Ethos, and the Civilian Creed.

Considerations when Choosing Techniques for Influencing

10-18. Advisors clearly understand what their counterparts value and fear. This information allows advisors to quickly relate the objective of the influence to something that their counterparts treasure. Also, a strong sense of self supports the advisor's recognition of which techniques they can successfully execute. Someone with an intimidating presence may achieve compliance with pressure more quickly than someone without that presence. Understanding the context of the request is important. For instance, using participation in the midst of flying bullets would not make sense. However, a long-term project that impacts many people might be a worthy situation for participation. Advisors also consider—

- The motivation of their counterparts.
- The relationship they have with their counterparts.
- What they do or not have to offer.
- An understanding of the second- and third-order effects their recommendations will have.
- What has worked in the past.
- Possession of an appropriate power based for use of a technique in particular context.
- Consistency of technique with prevailing social norms and role expectations.
- Level of resistance encountered or anticipated.
- Costs of using a technique in relation to likely benefits.

10-19. The cost of a technique refers to the amount of effort, favor, or resources that accompany the technique. A personal appeal that relies on friendship is quite effective; however, counterparts might expect the favor to be returned. If resources or time are plentiful, then the exchange might not seem expensive, but if they are not plentiful, then advisors must choose a different technique. Advisors balance the costs and benefits of each technique.

Moderating Factors

10-20. Sometimes moderating factors diminish the impact of any influence technique. These factors may include cultural misunderstanding, level of trust, or constraints on the counterpart like threats or corruption. Advisors analyze and learn about these factors. Advisors must be flexible when using influencing techniques. When one technique doesn't work, switching strategies or adding techniques may be appropriate.

CONSIDERATIONS

10-21. Seven considerations need to be addressed to influence behavior. These include determining the goal; who to influence; motives; beliefs, values, and attitudes; cultural beliefs, values, and attitudes; and susceptibility. These techniques are designed to help advisors achieve their goals. Just knowing that the counterpart wants something is not enough. Whether it is time off, new weapons, or money, advisors have a limited supply of stuff they can promise; they can only bend so much on rules, regulations, and laws. It is at this point that the advisor enters the realm of negotiations.

Determine the Goal

10-22. Advisors influence others to achieve some purpose. To be successful at exerting influence, advisors have an end state or goal in mind. Sometimes the goal will be very specific. Many goals are less distinct and measurable but are still valid and meaningful.

Determine Who to Influence

10-23. Advisors develop a mental list of those whom they might need to influence. Getting a FSF platoon to take action may be as simple as influencing the commander, but the advisor may have to influence the soldiers and commander of the platoon. When practical, advisors consider many variations of whom to influence. They look for circumstances in the immediate environment that significantly affect their counterparts. Individual or group characteristics to consider can include—

- Gender.
- Religion.
- Age.
- Socioeconomic status.
- Ethnicity.
- Political affiliation.

- Level of education (important in determining how to access and persuade).
- Occupation.

10-24. The challenge is to determine which characteristics impact emotional or behavioral responses and under what circumstances. Ethnicity may significantly affect behavior and emotions regarding voting and politics but may have almost no effect on the decision to enlist in the military. Individual and group characteristics may provide additional insight on motivations.

Determine and Prioritze Motives

10-25. Motives are anything that is required or desired (needs and wants). Motives are the reasons behind the decisions people make. Motives come from an inner desire to meet a need or want. Advisors use the knowledge of what motivates others to influence them. Knowing one's counterpart and others who may be influenced gives advisors insight into influencing their behavior. The desires to fulfill, alleviate, or eliminate a need or want provides the motivation to change behavior.

Basic Motives

10-26. Basic motives involve physiological needs—food, water, and air—and safety needs—security and shelter. When people feel hunger, they are motivated to eat; when people feel pain, they are motivated to reduce the source of the pain. Such basic motives are extremely powerful in driving behavior and overwhelm psychological needs and wants.

Social Motives

10-27. Social motives associate access to other tangibles (money, goods, education, infrastructure, and healthcare) or more complex psychological motives with basic motives. For example, people learn to want money because it can be exchanged for food and other desired goods. Power, achievement, reassurance, escape, justice, and acceptance illustrate types of learned social motives. Examples include:

- Wanting better educational opportunities for their children.
- Wanting better paying jobs.
- Wanting their interests represented by the government.
- Wanting revenge for perceived wrongs.
- Wanting self rule.

Prioritize Motives

10-28. Advisors prioritize motives by immediacy of the need or want. Critical motives are immediate needs. Short-term motives are currently being satisfied or active efforts are being made to satisfy them in the near future. Long-term motives are desired but are not immediately important; satisfaction may be delayed until some point in the future.

Determine Beliefs, Values, and Attitudes

10-29. Although often difficult to derive, beliefs, values, and attitudes can prove quite effective in persuasion. Beliefs are what is thought to be true. People do not believe in something if they think it is not true. People vary from being gullible to skeptical. Values prioritize what one believes. Values can be ethical, doctrinal, social, aesthetic, or economic in nature. Attitudes reflect whether one likes or dislikes something. Likes and dislikes can often be traced back to a person's beliefs and values. This includes loves, hates, frustrations, and fears. Frustrations occur when a want is not met. By examining causes of frustration, the advisor may find another motive to address.

Compare Beliefs, Values, and Attitudes

10-30. The predominant cultural beliefs, values, and attitudes; traditions; and norms may differ greatly from the group the advisor is trying to influence. Many Americans act differently in different cultures. This change in behavior relates directly to their predominant culture.

Determine Susceptibility

10-31. Susceptibility is the likelihood that the individual or group will be open to persuasion. Determining susceptibility aids in prioritizing efforts. If susceptibility is moderate to high, then advisors may not need to focus as much effort as they would if the susceptibility is low. Advisors determine susceptibility by assessing the strength of the underlying source of the motivation and the perception of the desired corresponding behavior. The strength of the motivation depends on the perceived risk and the perceived reward, while the perception of the behavior directly links directly to how consistent the behavior is with the individual or groups' values and beliefs.

Perceived Risks and Rewards

10-32. Risks may include the threat of physical harm, financial loss, or disapproval of peers. Rewards may include financial gain, political power, approval of peers, or increased self-esteem. Risks and rewards can vary from immediate- to long-term. Immediate risks and rewards are stronger than long-term risks and rewards. For example, few people will walk a tightrope (immediate risk), but many people smoke (long-term risk). The higher the perceived rewards and the lower the perceived risks, the more likely the behavior will occur.

Consistency with Values and Beliefs

10-33. Advisors determine if the desired behavior is consistent with values and beliefs. Generally, behavior is consistent with the relative moral or ethical, religious, political, and cultural beliefs, values, and norms. Behavior that is inconsistent or incongruent with values and beliefs often decreases over time. Alternately, if behavior continues to be incongruent with values and beliefs, then individuals are likely to justify this behavior by modifying their values and beliefs. The more consistent a desired behavior change is with the individual's values and beliefs, the more likely the behavior will occur.

10-34. The lower the susceptibility, the less likely influencing behavior will be successful, which does not mean that the person cannot still be influenced. The lower the susceptibility, the more conditions will need to be modified for behavior change to occur.

Application

10-35. Influencing is a dynamic process. Advisors make a plan before attempting to influence. If one technique fails, they have to be able to shift to other techniques. Advisors do not need to sit down and write a plan before attempting to influence people. However, even with apparently positive results, advisors review their tactics and techniques.

Checks

10-36. Advisors check measures of effectiveness, unintended consequences, and second- and third-order effects. Advisors assess if their influencing is achieving their immediate measures of effectiveness. However, with every action some unintended consequences may occur or the activity may not prepare counterparts for more complicated tasks later on.

CROSS-CULTURAL NEGOTIATING

10-37. The principal form of negotiation is likely to be cross-cultural negotiation. Cross-cultural negotiation brings with it a series of unique challenges and techniques. Culture fundamentally affects language and behavior. It also significantly impacts the way people handle conflict.

10-38. Negotiations among members of the same culture can be stressful; negotiating with members of other cultures can be exceedingly difficult. Logically, working with other cultures is a basic skill for the Soldier and an absolute requirement when acting in an advisory role. Understanding the components, perspectives, roles, and outcomes will help minimize the difficulty of cross-cultural negotiation. Chapter 4 of TC 31-73 discusses cross-cultural negotiations.

INFLUENCES ON NEGOTIATIONS

10-39. Cultural differences may lead to a conflict between what the parties expect and what their families and communities expect. This has an obvious impact on negotiating behavior. Negotiations are influenced by perceptions, individualism, punctuality and pace, relationship building, and language.

Perceptions

10-40. Most Americans have certain preconceived ideas about people from other cultures. These perceptions may not be factually based, but they exist and influence negotiations. Similarly, foreign negotiators have certain perceptions about Americans. Again, these perceptions may not be based on fact, but they exist. Advisors need to understand how other cultures perceive Americans to appropriately adjust their negotiating style. Successful advisors find ways to capitalize on the positive preconceptions of their counterparts and find ways to neutralize the negative preconceptions.

Individualism

10-41. The United States encourages a largely individualistic culture. As such, the typical American negotiator prefers to enter into negotiations alone or, if required, as part of a very small negotiating team (two or three people, at most). This behavior is largely a by-product of U.S. culture, which focuses on the individual, performance, initiative, and accomplishments. Most other cultures of the world place less emphasis on the individual and more on the group.

Punctuality and Pace

10-42. Americans place a great emphasis on punctuality, viewing it as an indication of an individual's basic ability and commitment. However, other cultures emphasize punctuality differently. Some emphasize punctuality more than Americans. In other cultures, time is considered relative or may not be important at all.

Relationship Building

10-43. Americans are very competitive during negotiations and are inclined to stress short-term results. Interpersonal relationships are of minimal importance to American negotiators. Building long-term relationships only occurs after the successful completion of negotiations.

Language

10-44. U.S. negotiators must be considerate of others' with respect to language. Even if the counterpart is fluent in English, they may expect negotiators to speak their language or use an interpreter. Members of some cultures do not like to speak English as they are very proud of their language. When U.S. negotiator counterparts do not speak English, they will need an interpreter. Counterparts may also expect American negotiators to present detailed written proposals in the local language. Many cultures appreciate foreigners who take the trouble to learn a few simple courtesy words. However, some prefer that foreigners who are not fluent refrain from using more than a few, basic courtesy words.

ELEMENTS FOR CONDUCTING CROSS-CULTURAL NEGOTIATIONS

10-45. The basic elements common to all negotiations must be applied in ways that allow for the differing cultures of the participants. Elements that must be considered include opening strategies, directness, strengthening behaviors, movement, power, face-saving, formal agreements, and mediation.

COMPONENTS OF NEGOTIATING

10-46. Two opposite types of negotiating exist: distributive and integrative. They reflect a philosophy towards negotiating in general. Distributive negotiating is win-lose while integrative is win-win. Distributive negotiating involves dividing a fixed amount of a thing to the parties involved. This negotiation normally involves people who have never had a relationship before. Integrative negotiating involves joining two parties to achieve something together. Successful negotiations can involve a mix of both strategies.

10-47. When using distributive negotiating, advisors know the point at which counterparts will walk away. Advisors push them almost to that point by manipulation and withholding information. The distributive negotiator believes—

- The disputants are adversaries and the goal is victory.
- Fixed resources are divided so one side receives more than the other after making concessions.
- One person's interests or position opposes the other's interests or position.
- The dominant concern is usually maximizing one's own interests.

10-48. Integrative strategies used by advisors include cooperation, sharing information, and mutual problem solving. The integrative negotiator believes—

- The disputants are joint problem solvers seeking a wise decision.
- An understanding of needs will allow resources to be divided and both sides can win.
- The relationship between the parties is what is critical.
- The dominant concern is to maximize joint outcomes.

Perspectives Affecting Negotiations

10-49. Four perspectives impact negotiations: time, environment, self, and nonverbal communication. Advisors consider how each party views these conditions when negotiating.

Time

10-50. Some cultures and individuals see time as a micro view. It is all about here and now. Other cultures and individuals see time in a macro view. Now is a point that bridges the past with the future. For example, imagine a person with the first scheduled event has a deadline to make an agreement now. The next scheduled event is 1500 so that person has two hours to come to an agreement. The second person might agree that they can come to an agreement now; but to the second person, now could be within the next month or two.

Environment

10-51. Some cultures believe they can impact, if not control, their environment. If something should go wrong, somebody is to blame. Every accident is preventable. Other cultures believe the environment impacts the person. Accidents just happen and no one could have done anything about it. One side may be quite ready to commit to something in the future *knowing* that nothing stands in the way, while the other side may be reluctant to talk in certainties of the future *knowing* that factors beyond their control may prevent them from fulfilling their commitments.

Self

10-52. In some cultures, when a person makes a mistake, blame can be isolated to that person. In other cultures, the individual is seen as an element of the group. So a mistake is made not by the individual but rather the whole group. In negotiations, one might accept a personal slight to get the goal achieved, whereas the counterpart represents the group and cannot allow shame to fall on them even though one might accept it personally.

Nonverbal Communication

10-53. Since advisors will often have to work through interpreters, nonverbal communication becomes more important. Some cultures respect people who mean just what they say and maintain their composure. Other cultures respect the eloquence in which something is said or the emotion used to convey it. Where words may be translated improperly, the advisor or counterpart can hear tone and see emotion.

Roles of Negotiation

10-54. When entering a negotiation, advisors must know what their role. The three major roles associated with negotiation are negotiator, mediator, and arbitrator. Each role has its own tactics or techniques and separate functions.

Negotiator

10-55. Advisors most often find themselves in the role of a negotiator. A negotiation is the process by which parties exchange commitments or promises to resolve disagreements and reach a settlement. Advisors prepare for each negotiation so they will not be at a disadvantage during the negotiation. To prepare for negotiations advisors examine their options, discover what motivates the parties, list all positive scenarios, and dispute resolutions.

10-56. Advisors examine their options. They question their goals, the importance of the goals, and the amount of leverage they have. They determine who is in power and what to do if no agreement is reached. Advisors uncover what motivates their counterparts. They delve into their counterparts' underlying needs.

10-57. By listing all the scenarios that satisfy their needs, advisors can maximize the gain for all parties. Advisors can add features to the scenarios that increase the opportunities for all parties to be satisfied. When advisors dispute resolutions, they often find some form of compromise. One of the best outcomes of compromise is when all parties leave the bargaining table satisfied but no single party feels like they have won or lost the argument. A skilled negotiator should strive to achieve a win-win result.

10-58. The negotiator represents one side and has a vested interest in the outcome of the negotiation. It is best to avoid situations where there is a conflict of interests. By blindly pushing the coalition agenda, the advisor could lose the trust of the counterpart. Considerations for a negotiator should include:

- If the foreign security forces ask the advisor to negotiate for them, they expect the advisor, as their negotiator, to have their best interests in mind. The coalition may try to use the advisor to push their agenda and influence rather than negotiate.
- If the foreign counterparts take illegal action, advisors cannot use commitment techniques. They need to use some form of compliant technique.
- If the foreign counterpart is not violating any laws, the advisor should focus on commitment techniques. Success will not be achieved all at once.

Mediator

10-59. Advisors may find themselves in the role of a mediator. The mediator has three general roles to perform: facilitator, formulator, and manipulator. Table 10-3 identifies the roles and tactics advisors use.

Table 10-3. Mediator roles and tactics

Roles	*Tactics*
Facilitator	• Make contact. • Arrange for interaction. • Clarify the situation. • Supply missing information. • Transmit messages between parties. • Ensure all interests are discussed.
Formulator	• Control the pace, formality, and physical environment of the meetings. • Highlight common interests. • Help parties save face. • Keep process focused on the issues. • Make substitutive proposals. • Suggest concessions parties could make.
Manipulator	• Keep parties at the table. • Change parties expectations. • Highlight costs of nonagreement. • Reward concessions. • Promise resources. • Threaten the withdrawal of resources. • Offer to verify compliance. • Threaten to withdraw mediation.

Arbitrator

10-60. The foreign security force may try to put the advisor in the position of an arbitrator. An arbitration is the process by which the parties to a dispute submit their differences to the judgment of an impartial person or group selected by mutual consent or statutory provision. They want the advisor to decide between two positions. This is a dangerous position for the advisor. When a win-win solution can be found, the advisor can improve rapport with both parties, but when it is a win-lose situation, the advisor could alienate one or both groups.

Expected Outcomes of Negotiation

10-61. The outcome of negotiations sometimes can determine the outcome. Advisors consider the desired outcome to determine which confrontation preference to use. Negotiation can focuses on physical outcomes of relationships. How advisors combine their preferences to their focus affects their perceptions of the outcome.

Confrontation Preferences

10-62. Negotiators will have one of three preferences for confrontation: avoidance, collaboration, or competition. Those that prefer avoidance will consider accommodation as a viable means to resolve an issue. Negotiators that prefer collaboration consider compromise as the preferred outcome. Those negotiators who use competition want to win. Intentionally losing in a negotiation to avoid confrontation seems counterintuitive; however, it tends to protect the current relationship. Collaboration requires a mutual respect of the existing relationship, while competition is generally indifferent. It would be fairly easy to take advantage of a negotiator who prefers to avoid confrontation. At the same time, expecting to be able to compromise with a highly competitive negotiator may become a point of frustration. For the competitor, winning may be more important than the deal itself.

Negotiation Focus

10-63. Defining whether the negotiation is a win-win situation is not as simple as which side gained the most from the negotiation. Advisors often focus on physical outcomes whereas diplomats focus on relationships. Every negotiation varies on which factor is more important. Just as car dealers may accept less profit on individual services, they gain overall if the customer returns.

Perception of the Outcome

10-64. Comparing confrontation preferences leads to nine potential combinations. If advisors can determine a participant's preference—avoidance, collaboration, and competition—and focus, they can determine how the participant might feel based on the outcome of the negotiation. If one participant is a collaborator and one is a competitor, then a win-win situation is possible, but neither side will be completely satisfied with the outcome. The competitors will not have gotten a clear win, and the collaborators will not have gotten what they considered fair. Subsequently, they will both have some minor negative feelings toward the other.

Appendix A
Army Special Operations Forces Imperatives

This appendix outlines the imperatives for Army special operations forces conducting security force assistance and foreign internal defense operations. Table A-1 lists these imperatives.

Table A-1. Army special operations forces imperatives

Imperatives	*General Descriptors*	*Specific Actions*
Understand the Operational Environment	• Knowing the U.S. command relationships (military and interagency) • Knowing the population (who, what, when, where, why, and how) • Knowing the enemy and other antigovernment forces (crime, militias, gangs) • Knowing the higher commander's intent • Knowing that the press is a part of the operational environment (news reports of events reach the U.S. population quickly)	• Define relationship as partnership or augmentation • Conduct analysis using ASCOPE (areas, structures, capabilities, organizations, people, events) • Analyze the enemy order of battle • Protect the population
Recognize Political Implications	• Identifying the political (and legal) restrictions and implications of actions by U.S. and foreign security forces • Identifying political implications of establishing and enforcing rules of engagement policy • Identifying short and long term political implications of security force assistance • Understanding the importance of establishing human rights (critical for tactical and strategic success) • Acknowledging the effects of public opinion on the mission locally, in the United States, and internationally • Understanding that insurgents conduct operations to have political effects among the local populace, external actors, and the U.S. and multinational civilian public	• Include rules of engagement in the planning and briefs of operations • Include local government leadership in operations • Reinforce human rights briefings and training to all foreign security forces • Prepare all U.S. and foreign security forces to talk with the press • Prepare information engagements for all operations • Anticipate and neutralize the methods enemy forces pass propaganda

Imperatives	*General Descriptors*	*Specific Actions*
Facilitate Interagency Activities	• Conducting security force assistance as both a military and an interagency effort; taking advantage of the greater impact of collective efforts • Coordinating the scope and limitations of each agency's influence and programs • (as opposed to a purely military program) • Obtaining the active cooperation of the foreign security forces • Including military and civilian agencies working together for unity of effort	• Include interagency members in planning and training as appropriate • Provide assistance to support limited agency resources, allowing agencies to prioritize their primary roles • Conduct regular coordination meetings with civil agencies to synchronize security efforts • Include host-nation agencies whenever including U.S. agencies
Engage the Threat Discriminately	• Avoiding the use of overwhelming firepower—to prevent alienating the population and turning them away from the host-nation government • Giving training and advice to foreign security forces based on the mission and threat • Targeting insurgents, criminal activities (lawlessness), or subversion—to avoid alienating the populace • Minimizing the use of force in populace and resources control measures—weighed for potential gain versus potential cost • Ensuring that rules of engagement and escalation of force are balanced between force protection and safeguarding civilians	• Take civilians into account when considering weapons systems and techniques for operations • Include civilians in risk assessment during mission planning • Train and rehearse rules of engagement and escalation of force to prepare Soldiers to react in accordance to policy • Publicize reasons for use of populace and resource control (for example, curfews protect civilians); avoid enemy propaganda or rumors dictating the psychological response of the populace • Update rules of engagement and escalation of force periodically • Rules of engagement or escalation of force reflects engaging targets that are visually verified and not simply suppressive fire in populated areas
Consider Long-Term Effects	• Building up the foreign security forces and alleviating the root causes of the current situation over time—not just treating the symptoms • Solving problems in a broader political, military, social and economic context • Making policies, plans, and operations consistent with national and theater priorities and the objectives they support	• Understand local grievances and underlying issues • Interact with all aspects of the population a government to identify root causes • Ensure the commander's intent remains focused on higher goals and policy related to addressing root causes, and protecting the local population • Monitor and update rules of engagement as local government and police gain control of rule of law. Conduct operation that adhere to local laws as possible • Understand legal and political constraints (rules of engagement) to avoid strategic failure while achieving tactical success

Imperatives	*General Descriptors*	*Specific Actions*
Ensure Legitimacy of Special Operations	• Reinforcing and enhancing the legitimacy and credibility of the foreign security forces and government • Gaining trust and credibility to ensure advisors and U.S. forces obtain legitimacy with the local population • Obtaining legitimacy in theater, in the United States, and internationally to ensure ongoing support	• Conduct joint and multinational patrols and operations. • Ensure information engagement is common in all operations and contributes to maintaining credibility • Engage the media with the host nation to inform as well as to justify actions • Focus on what the population considers the root causes, not the U.S. perspective • Develop long-term solutions to restore or improve basic services with host-nation leadership
Anticipate and Control Psychological Effects	• Considering the psychological effects of all operations (such as populace and resources control measures, combat operations, and civic action) • Avoiding negative psychological effects that can negate tactical victories • Appreciating the power of people's perceptions versus reality	• All operations are developed, analyzed and executed with information engagement • Continue to gauge the psychological effects with the local population, local government, and security forces as well as the enemy (propaganda and interrogation)
Apply Capabilities Indirectly	• Teaching host-nation forces to plan instead of giving them tasks • Supporting U.S. efforts enhancing the legitimacy and credibility of the host-nation government and foreign security forces • Balancing control and pressure on foreign security forces to ensure they conduct successful tactical operations and accomplish missions • Fulfilling primary role to advise, train, and help foreign security forces take the lead and stabilize the host nation • Influencing a foreign security force without commanding it; not accepting a lack of effort or a high level of corruption	• Conduct training of foreign security forces only to teach basics • Transition to mentor or advisor relationship during planning and operations to encourage the host-nation to take the lead • Use discretion when advising counterparts as not to embarrass foreign security forces leadership • Use host-nation methods whenever possible

Imperatives	*General Descriptors*	*Specific Actions*
Develop Multiple Options	• Considering the complex and irregular environment • Planning to use the full range of capabilities • Anticipating that an operational environment may dictate a change of rules of engagement or type of operations • Ensuring foreign security forces are capable of shifting from one option to another before and during mission execution	• Continue to refine and train on battle drills and contingencies • Develop an understanding of skills and abilities of the members of the unit and community • Elicit the input of foreign security forces on how to address situations to widen the options beyond conventional techniques • Include government agencies, special operations forces, nongovernmental organizations, local police, and the host-nation government to develop additional methods or options for solving problems
Ensure Long-Term Sustainment	• Avoiding advising and training foreign security forces in techniques and procedures beyond their capabilities to sustain • Modifying tactics, techniques, and procedures; training; operations; and sustainment to fit the culture, education, and technological capability of the foreign security forces • Developing host-nation sustainment before handing over economic, social, political, military, and security initiatives and projects • Avoiding host-nation dependency on U.S.-funded programs	• Understand and utilize the equipment and manpower present in the foreign security forces unit's inventory • Assess the capabilities of foreign security forces for leadership, training, sustainment (logistics, administration, and equipment), and ongoing development of a professional security force • Recognize the programs that are durable, consistent, and sustainable by the host nation • Do not begin programs beyond the economic or technological capacity of the host nation
Provide Sufficient Intelligence	• Ensuring detailed, near real-time, all-source tactical intelligence products support all activities and programs • Ensuring detailed and comprehensive intelligence on all aspects of the operational environment and its internal dynamics support operations • Ensuring human intelligence, urban recon, and internal security elements assess internal threats, warn the government, and take action to penetrate and neutralize instability • Considering aspects of the society not directly related to the tactical combat situation when assessing the enemy threat • Using information from intelligence assets to advise, train, and help foreign security forces counterparts and ease interagency efforts	• Ensure host nation forces are conducting operations using accurate, real-time intelligence • Establish priority of effort when identifying intelligence requirements • Assist foreign security forces to prioritize and focus counterintelligence on the commander's critical information requirement to receive timely and specific intelligence • Assess other organizations that impede the rule of law and government functions (such as gangs, criminals, smugglers, militias, mercenaries, drug lords, and narcoterrorists)

Imperatives	General Descriptors	Specific Actions
Balance Security and Synchronization	• Maintaining unity of effort and avoiding compartmentalizing key security forces from planning and intelligence contingencies even when security concerns exist • Avoiding either excessive or insufficient security • Anticipating reasons foreign security forces may not provide intelligence, such as lack of internal security or lack of U.S. intelligence support	• Encourage combined planning, preparation, and operations with foreign security forces' military, police, and governmental units • Resolve conflicting demands on mission planning and execution • Employ measures to mitigate risk of operations security instead of dismissing the host-nation government, police, or military needed for the overall effort • Strive to gain the trust of the local population to gain the most human intelligence and support from the foreign security forces

Appendix B
Legal Considerations

GENERAL LEGAL BASIS

B-1. Law and policy govern the actions of U.S. forces in all military operations, including security force assistance. For U.S. forces to conduct operations, a legal basis must exist. This legal basis profoundly influences many aspects of an operation. It affects the rules of engagement, how U.S. forces organize and train foreign forces, the authority to spend funds to benefit the host nation, and the authority of U.S. forces to detain and interrogate. Under the Constitution of the United States, the President is the Commander in Chief of U.S. forces. Therefore, orders issued by the President or the Secretary of Defense to a combatant commander provide the starting point in determining the legal basis. Laws are legislation passed by Congress and signed into law by the President and treaties to which the United States is party. Policies are executive orders, departmental directives and regulations, and other authoritative statements issued by government officials. This appendix summarizes some of the laws and policies that bear upon U.S. military operations in support of security force assistance. This summary does not replace a consultation with the unit's supporting staff judge advocate.

LEGAL AUTHORITY FOR SECURITY FORCE ASSISTANCE

B-2. If the Secretary of State requests and the Secretary of Defense approves, U.S. forces can participate in security force assistance. The request and approval can go through standing statutory authorities in Title 22, U.S. Code. Title 22 contains the Foreign Assistance Act of 1961, the Arms Export Control Act of 1976, and other laws. It authorizes security assistance, developmental assistance, security force assistance and other forms of bilateral aid.

B-3. The request and approval for security force assistance might also occur under various provisions in Title 10, U.S. Code. Title 10 authorizes certain types of military-to-military contacts, exchanges, exercises, and limited forms of humanitarian and civic assistance in coordination with the U.S. ambassador to the host nation. In such situations, U.S. forces may be granted status as administrative and technical personnel based on a status-of-forces agreement or an exchange of letters with the host nation. This cooperation and assistance is limited to liaison, contacts, training, equipping, and providing defense articles and services. It does not include direct involvement in operations. Assistance to police by U.S. forces is permitted but, generally, Department of Defense (DOD) does not serve as the lead government department. Without receiving a deployment or execution order from the President or Secretary of Defense, U.S. forces may be authorized to make only limited contributions during operations that involve security assistance.

EXISTING UNITED STATES LAW

B-4. The Constitution divides the power to wage war between the executive and legislative branches of government. Under Article I, Congress holds the power to declare war; to raise and support armies; to provide and maintain a navy; and to make all laws necessary and proper for carrying out those responsibilities. Balancing that legislative empowerment, Article II vests the executive power in the President and makes the President the Commander in Chief of the Armed Forces. This bifurcation of the war powers created an area in which the coordinate political branches of government exercise concurrent authority over decisions relating to the use of Armed Forces overseas as an instrument of U.S. foreign policy.

B-5. The Supremacy Clause of the Constitution (Article VI) states, in part, that all treaties made by the United States are the supreme law of the land. Therefore, ratified treaties, such as the United Nations (U.N.) Charter (see paragraphs B-7 through B-8) and the Geneva Conventions, create legal obligations on U.S. forces regarding their ability to perform various types of missions and functions.

B-6. The War Powers Resolution of 1973 ensures the collective judgment of both the executive and legislative branches to deploy U.S. forces by requiring consultation of and reports to Congress, in any of the following circumstances:

- Introduction of troops into actual hostilities.
- Introduction of troops, equipped for combat, into a foreign country.
- Greatly enlarging the number of troops, equipped for combat, in a foreign country.

B-7. The President is required to make such reports within 48 hours of the triggering event, detailing—

- The circumstances necessitating introduction or enlargement of troops.
- The constitutional or legislative authority upon which he bases his action.
- The estimated scope and duration of the deployment or combat action.

B-8. Since the War Powers Resolution was passed over the veto of President Nixon, no President has either conceded its constitutionality or complied fully with its mandates.

International Law and Treaties

B-9. The U.N. Charter became effective on 24 October 1945 after being ratified by the United States and a majority of other signatories. The U.N. Charter mandates that all member states resolve their international disputes peacefully. It requires that member states refrain in their international relations from the threat or use of force. The U.N. Charter also provides that all nations have the right to use self-defense to combat acts of aggression against them until such time as the Security Council takes action.

B-10. The U.N. Charter provides the essential framework of authority for the use of force, effectively defining the foundations for a modern *jus ad bellum.* Inherent in its principles are the requirements for necessity, proportionality, and timeliness. Necessity involves considering—

- The exhaustion or ineffectiveness of peaceful means of resolution.
- The nature of coercion applied by the aggressor State.
- The objectives of each party.
- The likelihood of effective community intervention.

B-11. Proportionality refers to limiting force in magnitude, scope, and duration to that which is reasonably necessary to counter a threat or attack. Timeliness refers to situations in which the delay of a response to an attack or the threat of attack attenuates the immediacy of the threat and the necessity to use force in self-defense.

B-12. U.S. forces obey the law of war. The law of war is a body of international treaties and customs, recognized by the United States as binding. It regulates the conduct of hostilities and protects noncombatants. The main law-of-war protections come from The Hague and Geneva Conventions. They apply at the tactical and operational levels and are summarized in ten rules shown in table B-1.

Table B-1. Summary of the law-of-war rules

• Fight only enemy combatants.
• Do not harm enemies who surrender; disarm them and turn them over to the chain of command.
• Do not kill or torture detainees.
• Collect and care for the wounded, whether friend or foe.
• Do not attack medical personnel, facilities, or equipment.
• Destroy no more than the mission requires.
• Treat all civilians humanely.
• Do not steal; respect private property and possessions.
• Do one's best to prevent violations of the law of war.
• Report all violations of the law of war to superiors.

B-13. During security force assistance operations, commanders must be aware of Common Article 3 of the Geneva Conventions and the status of insurgents under the laws of the host nation. One article contained in all four of the Geneva Conventions—Common Article 3—is specifically intended to apply to internal armed conflicts:

> *In the case of armed conflict not of an international character occurring in the territory of one of the High Contracting Parties, each Party to the conflict shall be bound to apply, as a minimum, the following provisions:*
>
> *(1) Persons taking no active part in the hostilities, including members of armed forces who have laid down their arms and those placed "hors de combat" by sickness, wounds, detention, or any other cause, shall in all circumstances be treated humanely, without any adverse distinction founded on race, colour, religion or faith, sex, birth or wealth, or any other similar criteria. To this end, the following acts are and shall remain prohibited at any time and in any place whatsoever with respect to the above-mentioned persons: (a) Violence to life and person, in particular murder of all kinds, mutilation, cruel treatment and torture; (b) Taking of hostages; (c) Outrages upon personal dignity, in particular humiliating and degrading treatment; (d) The passing of sentences and the carrying out of executions without previous judgment pronounced by a regularly constituted court, affording all the judicial guarantees which are recognized as indispensable by civilized peoples.*
>
> *(2) The wounded and sick shall be collected and cared for. An impartial humanitarian body, such as the International Committee of the Red Cross, may offer its services to the Parties to the conflict. The Parties to the conflict should further endeavor to bring into force, by means of special agreements, all or part of the other provisions of the present Convention. The application of the preceding provisions shall not affect the legal status of the Parties to the Conflict.*

HOST-NATION LAW AND STATUS-OF-FORCES AGREEMENTS

B-14. After considering the type of baseline protections represented by fundamental human rights law, the military leader must be advised in regard to the other bodies of law that leader should integrate into planning and execution. This includes consideration of host-nation law. Because of the nature of most noninternational armed conflict missions, commanders and staffs must understand the technical and pragmatic significance of host-nation law within the area of operations. Although in theory, understanding the application of host-nation law during military operations is perhaps the simplest component, in practice it is perhaps the most difficult.

B-15. Judge advocates must recognize the difference between understanding the technical applicability of host-nation law and that law's application to control the conduct of U.S. forces during operations. In short, the significance of host-nation law declines in proportion to the movement of the operation toward violent conflict. Judge advocates should understand that U.S. forces enter other nations with a legal status that exists anywhere along a notional legal spectrum.

B-16. Judge advocates and planners must be familiar with any status-of-forces agreements or status-of-mission agreements that may be applicable. In any given country, there may be agreements short of status-of-forces agreements, such as diplomatic notes, on point. Relevant international documents affecting military operations may be difficult to locate. Several sources are available for locating applicable international agreements governing the status of U.S. forces or affecting military operations. Department of State (DOS) publications, such as *Treaties in Force,* contain unclassified international agreements. Both the relevant geographic combatant command's legal office and the defense attaché or military assistance group at the embassy normally have access to host-nation or international agreements impacting a military operation.

B-17. Status-of-forces and other forms of agreements frequently exist. They are essentially contractual agreements or treaties between two or more nations that establish the legal status of military personnel in foreign countries. Topics that are usually covered in a status-of-forces agreement include criminal and civil jurisdiction, taxation, and claims for damages and injuries. In the absence of an agreement or some other arrangement with the host nation, DOD personnel in foreign countries may be subject to its laws. Commanders ensure that all personnel understand status of U.S. forces in the area of operations and are trained accordingly.

LEGAL CONSTRAINTS ON THE SECURITY FORCE ASSISTANCE MISSIONS

B-18. U.S. law and regulation play a key role in establishing the parameters by which military forces may conduct security force assistance missions. These parameters tend to constitute constraints on the activities of military units. They range from the rules of engagement in combat situations to the ability to spend government funds in furtherance of a training or support mission.

GENERAL PROHIBITION ON ASSISTANCE TO POLICE

B-19. Historically, DOD is not the lead government department for assisting foreign governments. DOS is the lead when U.S. forces provide security assistance—military training, equipment, and defense articles and services—to host-nation governments. The Foreign Assistance Act of 1961 specifically prohibits assistance to foreign police forces except within specific exceptions and under a Presidential directive. When providing assistance to training, DOS provides the lead role in police assistance through its Bureau of International Narcotics and Law Enforcement Affairs. The President, however, may delegate this role to other agencies. This was done in 2004, when the President signed a decision directive granting the commander, United States Central Command, authority to train and equip Iraqi police.

TRAINING AND EQUIPPING FOREIGN SECURITY FORCES

B-20. All training and equipping of foreign security forces are specifically authorized. U.S. laws require Congress to authorize expenditures for training and equipping foreign forces. The laws of the United States also require DOS to verify that the host nation receiving the assistance is not in violation of human rights. The Secretary of Defense may authorize deployed U.S. forces to train or advise host-nation security forces as part of the mission. In this case, DOD personnel, operations, and maintenance appropriations provide an incidental benefit to those security forces. Numerous other programs to assist foreign forces are paid for with funds appropriated by Congress for that purpose. Consultation with a staff judge advocate or legal advisor early in the planning ensures the availability of funds for missions to train and equip foreign forces.

RULES OF ENGAGEMENT

B-21. *Rules of engagement* are directives issued by competent military authority that delineate the circumstances and limitations under which United States forces will initiate and/or continue combat engagement with other forces encountered (JP 1-02). Often these directives are specific to the operation. If there are no operation-specific rules of engagement, U.S. forces apply standing rules of engagement. When operating with a multinational force, commanders must coordinate the rules of engagement thoroughly.

FISCAL LAW CONSIDERATIONS

B-22. In security force assistance missions, like all operations, commanders require specific authority to expend funds normally found in the annual DOD appropriations act. As a general rule, operation and maintenance funds may not be used for security force assistance missions. In recent years, Congress appropriated additional funds to commanders for conducting more complex stability operations that were not typically covered by operation and maintenance funds. Examples include the Commander's Emergency Response Program (CERP), the Iraq Relief and Reconstruction Fund, Iraq Freedom Fund, and Commander's Humanitarian Relief and Reconstruction.

B-23. DOS has the primary responsibility, authority, and funding to conduct foreign assistance on behalf of the U.S. government. Foreign assistance encompasses any and all assistance to a foreign nation, including security assistance (assistance to the internal police forces and military forces of the foreign nation), development assistance (assistance to the foreign government in projects that will assist the development of the foreign economy or their political institutions), and humanitarian assistance (direct assistance to the population of a foreign nation). The legal authority for DOS to conduct foreign assistance is found in the Foreign Assistance Act of 1961.

B-24. There are two exceptions to the general rule requiring the use of Title 22 funds for foreign assistance. The first exception is for interoperability, safety, and familiarization training. DOD may fund the training (as opposed to goods and services) of foreign militaries with operation and maintenance dollars only when the purpose of the training is to enhance interoperability, familiarization, and safety training. Operation and maintenance funds may not be used for security force assistance training. This exception applies only to interoperability training. The second exception is congressional appropriation or authorization to conduct security foreign assistance. This exception allows DOD to fund foreign assistance operations if Congress has provided a specific appropriation or authorization to execute the mission.

Constraints on Funding

B-25. The Foreign Operations Appropriation Act (an annual DOS appropriations act) was enacted in 1997. Known as the Leahy Amendment, it prohibits the U.S. government from providing funds to a unit of the security forces of a foreign country if DOS has credible evidence that the unit has committed gross violations of human rights. The Leahy Amendment contains additional constraints on government funding of security force assistance missions. The law restricts funding until the Secretary of State determines and reports that the government of such country is taking effective measures to bring the responsible members of the security forces unit to justice.

B-26. Congress specifically appropriates funds for foreign assistance. The United States Agency for International Development (USAID) expends such funds under the legal authorities in Title 22. Provisions of Title 10 may also authorize amounts of money for these purposes. Standing funding authorities are narrowly defined and generally require advance coordination within DOD and DOS.

B-27. Effective foreign forces need training and equipment. U.S. laws require Congress to authorize such expenditures. U.S. laws also require DOS to verify that the host nation receiving the assistance is not in violation of human rights.

Funding Authorization for Security Force Assistance

B-28. All training and equipping of foreign security forces must be specifically authorized. Military and civilian personnel, operations, and maintenance appropriations should typically provide only an incidental benefit to those security forces. All other weapons, training, equipment, logistic support, supplies, and services provided to foreign forces must be paid for with funds appropriated by Congress for that purpose. Examples include the Iraq Security Forces Fund and the Afghan Security Forces Fund. Moreover, the President must give specific authority to DOD for its role in training and equipping foreign security forces. In May of 2004, the President signed National Security Presidential Directive (NSPD) 36 that made the commander, U.S. Central Command, under policy guidance from the chief of mission, responsible for coordinating all U.S. government efforts to organize, train, and equip Iraqi Security Forces, including police. Absent such a directive, DOD lacks authority to take the lead in assisting a host nation to train and equip its security forces.

Key Security Assistance and Foreign Assistance Funding Programs

B-29. Congress has appropriated funds to allow DOS to conduct its foreign assistance mission, including the following activities:

- Foreign Military Financing Program.
- International Military Education and Training Program.
- Economic Support Fund.
- Peacekeeping operations.
- Antiterrorism assistance.
- Global humanitarian demining.

- Refugee assistance.
- Personnel details.

B-30. DOS administers the following programs but no funds are appropriated to sustain them:

- Foreign Military Sales Program.
- Foreign Military Lease Program.
- Economy Act Security Assistance.
- U.S. Government Commodities and Services Program.
- Direct Commercial Sales Program.

B-31. Additional special support created by law for the foreign assistance mission includes—

- Excess defense articles.
- Presidential drawdowns.

B-32. DOS directly, or indirectly through USAID, finances numerous development assistance activities to address the following needs:

- Rule of law.
- Governance.
- Agriculture and nutrition.
- Population control.
- Health.
- Education.
- Energy.
- Environment improvement.

B-33. CERP (see paragraph B-19) is an example of a targeted humanitarian assistance fund program. CERP's primary purpose is to enable military commanders to respond to urgent humanitarian relief and reconstruction requirements within their fields of interest by carrying out programs that will immediately assist the host-nation populace. CERP was originally funded with seized Iraqi assets, but Congress later appropriated U.S. funds for the purpose. CERP is not applicable to missions outside Iraq and Afghanistan. Future missions, though, may have similar funding sources established to facilitate a humanitarian assistance mission. Commanders consult with the servicing judge advocate to determine the availability of funds.

Appendix C
Information Engagement

INFORMATION SUPERIORITY

C-1. Information shapes a commander's operational environment at every level. Twenty-first century conflicts occur in operational environments with instant communication. Because information systems are everywhere and exposure to news and opinion media is pervasive, an individual's words and actions can have immediate operational and strategic implications. Gaining and maintaining the trust of key publics has never been more important. Army forces conduct operations in the public eye and must maintain public confidence by being consistent in word and deed.

C-2. *Information superiority* is the operational advantage derived from the ability to collect, process, and disseminate an uninterrupted flow of information while exploiting or denying an adversary's ability to do the same (JP 3-13). Advisors build partner capacity in this mission area. Advisors ensure that host-nation counterparts integrate information superiority into operations as carefully as fires, maneuver, protection, and sustainment. Army information tasks contribute to information superiority. Information tasks are used to shape the operational environment (see FM 3-0). The information tasks are—

- Information engagement.
- Command and control warfare.
- Information protection.
- Operations security.
- Military deception.

C-3. Advisors prioritize the five information tasks based on the host nation's capabilities and strengths. This appendix discusses some information engagement capabilities that advisors can expect to employ while conducting security force assistance, as well as considerations when working with the media. For a further discussion of the other information tasks, see FM 3-0.

INFORMATION ENGAGEMENT CAPABILITIES

C-4. *Information engagement* is the integrated employment of public affairs to inform U.S. and friendly audiences; psychological operations, combat camera, and U.S. government strategic communications and defense support to public diplomacy, and other means necessary to influence foreign audiences; and leader and Soldier engagements to support both efforts (FM 3-0).

C-5. In general, advisors focus on information engagement because land operations occur among civilian populations. Advisors can have a significant impact by focusing on this task. Advisors work with host-nation counterparts to communicate information to the populace, build their trust and confidence in friendly forces, promote popular support for friendly operations, and positively influence public perceptions and behavior. Advisors integrate information engagement throughout the operations process.

C-6. Advisors ensure host-nation counterparts focus their information engagement activities on achieving desired effects locally. However, land operations always occur in a regional and even global context. Therefore, local information engagement activities must support and complement higher headquarters information engagement activities, host-nation and U.S. government strategic communications guidance (when available), and broader host-nation and U.S. government policy where applicable.

C-7. Three information engagement capabilities deserve particular attention as the advisor can directly influence them. These are public affairs, combat camera, and leader and Soldier engagements.

PUBLIC AFFAIRS

C-8. First, public affairs can have a profound and lasting effect on public perceptions of operational success or failure. This effect may have strategic consequences, even when related to tactical operations. Therefore, advisors ensure that public affairs efforts are integrated at tactical, operational, and strategic levels. (See paragraphs C-10 through C-20 for more information to help advisors working directly with the media.)

COMBAT CAMERA

C-9. Second, combat camera is used to document a wide range of tactical and operational successes and to counter enemy propaganda. If there is no dedicated host-nation combat-camera capability, the advisor recommends that a host-nation soldier be designated, equipped, and trained for this function. If that is not possible, the advisor should obtain a digital camera and, if necessary, photograph relevant host-nation activities personally. Positive photographs can have a tremendous benefit, and digital photos are easily taken and transmitted.

LEADER AND SOLDIER ENGAGEMENTS

C-10. Third, leader and Soldier engagements are a fundamental and highly effective method to influence people. They involve face-to-face interaction by leaders and Soldiers to influence the perceptions of the local populace. Military leaders engage key communicators and civilian leaders (often referred to as key leaders) through carefully planned meetings. Examples of key leaders include local tribal elders, elected or appointed officials, military and police commanders, and nongovernmental organization leaders.

WORKING WITH THE MEDIA

C-11. The appropriate public affairs officer normally manages media contacts. However, this is not always possible. A deployed advisor team seldom has a public affairs officer representative.

C-12. Occasions may arise when the host-nation press wants to question advisors. Even if an advisor team lacks a public affairs officer representative, advisors must avoid refusing to speak to the media. Apparent refusal to speak with accredited members of the media, for whatever reason, may inadvertently create strong negative impressions with strategic implications. Therefore, advisor teams designate a media spokesperson and keep this individual well informed. This individual coordinates media contacts for the advisor team. This individual ensures that official, published public affairs officer guidance and command policies regarding media relations, including anonymity and operations security, are observed. The following paragraphs provide general media relations guidance for advisor teams.

ENHANCING RELATIONSHIPS WITH THE MEDIA

C-13. Advisors strive to convey a positive impression and enhance relationships between the media and the military. If deemed acceptable within the public affairs guidance and command policies, the advisor team's designated media spokesperson—

- Maintains a list of reliable reporters and editors that cover operations in the team's operational area and keeps them informed of significant activities.
- Answers media inquiries promptly, accurately, and courteously.
- Finds out reporters' deadlines and uses them to the friendly force's advantage.
- Stresses the human aspects of a story that readers, viewers, or listeners can identify with, including the negative impact of opposing operations on the people.
- Points out how advisors and host-nation counterparts are working to address the needs of the unfortunate.
- Encourages the media to observe constructive work the host nation and advisors are doing and to interview them about their efforts.
- Identifies a location in advance for the press to take photos and videos.

Correcting Public Misconceptions

C-14. If information is withheld, journalists often fall back on speculation. Such speculation is usually inaccurate. However, it is often near enough to the truth that it is easily accepted by large sections of the public and possibly by established governments. In addition, partisan sources may find it advantageous to leak part of a story to the press to build up public support for their position. On occasion, such activities can grow into fully orchestrated press campaigns.

C-15. With modern satellites and communications technology, media are able to distribute reports and photographs extremely quickly. Incidents, sometimes fabricated or slanted toward a partisan viewpoint, may be aired in living rooms across the globe the same day.

C-16. However, advisors and host-nation counterparts frequently can correct public misconceptions and provide accuracy, adequate context, and proper characterization promptly. They can turn a potential liability into a major advantage by remaining prepared to coordinate and communicate with the media. They can take advantage of the media's powerful and far-reaching ability to communicate with critical audiences.

C-17. If the media report inaccurate or damaging information, advisors use caution in deciding on a course of action. Advisors (usually through the team's designated media spokesperson) inform their chain of command. Advisors consider the following questions:

- Is it important enough to correct or is it a detail that—in the long term—is not really important?
- How damaging is the charge, criticism, or error? Will a correction simply give greater visibility to an unfavorable point of view?
- Is a correction worth a restatement of the entire problem to new audiences who did not read or see it the first time?
- Is it possible to target only the audience originally exposed to the story?
- Can significant gain result from pointing out an error?

C-18. When incorrect or misleading information is released to the public through host-nation government or media sources, the advisor team determines a course of action appropriate to the situation. It must include a plan to ensure that relevant host-nation government officials and media recognize the inaccuracy of the information. This may prevent future incidents. Operational and combat experience since 2001 shows that in the struggle for information superiority, friendly forces must be candid and honest. Additionally, they must not allow lies disseminated by enemies to stand unchallenged, particularly in areas of great international interest.

Interviews With Advisor Team Members

C-19. Advisor teams sometimes arrange for media interviews with selected advisor team members. Properly conducted interviews can support the mission and even improve morale. When preparing for an interview, the team's designated spokesperson—

- Provides all team members with a single and simple theme to convey to the press, sometimes in the form of a mission narrative (a succinct account of the mission as a whole, explained consistently with the host nation's cultural view).
- Conducts rehearsals in front of video cameras, using mock interviews to practice communication of the theme.
- Selects interviewees based on their understanding of the theme covered, not their comfort level in front of a reporter. (Shyness is a normal trait that may add a human touch to a sensitive situation.)

C-20. If an advisor team member does not know the answer to a reporter's question, the individual should attempt to find out and get back to the reporter or refer the reporter to an appropriate source. This may establish the team as a helpful information source and develop a relationship for future balanced coverage.

C-21. Communicating with the media is the most difficult aspect of information engagement. Anything advisors say or do publicly can impact the mission. Therefore, advisors ensure consistency with national policy, professionalism, and clear communication. Advisors must know and follow the policies of the geographic combatant commander, the American Embassy, and the appropriate public affairs officer regarding media interviews. Additionally, advisors follow the specific guidelines shown in table C-1.

Table C-1. Specific guidelines for media interviews

- Maintain a friendly, yet professional attitude.
- Remain positive, greet interviewers, and welcome questions. (This is difficult under stressful or tragic conditions; however, a calm, mature appearance earns respect from audiences.)
- Use language that is clear and easy to understand.
- Avoid military jargon or terminology that others can misinterpret.
- Never enter an interview unprepared.
- Prepare notes and study them carefully; use facts to back up statements.
- Ensure the facts supporting their position are up to date and come to mind easily.
- Anticipate possible questions and think about various responses.
- Rehearse before interviews on location if possible and, if time permits, role-play tough questions.
- Learn and use reporters' names.
- Research interviewers' organizations, including—
    - The organizations' views or views they defend.
    - Their underlying motivations.
    - How members of the organizations previously conducted interviews.
- Keep the interests of the local nationals or other beneficiaries of the mission at the forefront.
- Avoid talking from the point of view of U.S. interests.
- Finish answering one question before answering another if two or more questions are asked at the same time.
- Avoid speculation and give only factual information that can be verified.
- Use caution in quoting statistics as such data may be easily disputed or reinterpreted.
- Avoid repeating a question that contains incorrect information or inflammatory language (this may result in a misquotation).
- Strive to tell the truth.
- Avoid saying *no comment.* (It gives the appearance the advisor has something to hide.)
- If unsure of something, admit it rather than mistakenly lie.
- If questioned about another agency's activities, refer to that agency for comment.
- Do not speak for other organizations.
- Avoid exaggeration or claims that cannot be backed up.
- Refuse all off-the-record discussions.
- Expect any statement they make to be quoted publicly.
- Avoid criticizing other groups or organizations, and avoid committing information fratricide.
- Remind reporters that photographing recognizable dead Soldiers, charts, maps, supply depots, or electronic warfare assets is off limits.
- Remind reporters that their personal security is not a primary military concern.
- Remain objective and calm.
- Demonstrate conduct befitting a representative the U.S. government
- Never give personal opinions and beliefs.
- Avoid reacting emotionally when interviewers appear skeptical or hostile.
- Maintain focus on the mission.

Glossary

SECTION I – ACRONYMS AND ABBREVIATIONS

ABCA	American, British, Canadian, Australian, and New Zealand Armies Program
AR	Army regulation
BCT	brigade combat team
CCIR	commander's critical information requirement
CERP	Commander's Emergency Response Program
COA	course of action
COIST	company intelligence support team
DA	Department of the Army
DOD	Department of Defense
DODD	Department of Defense directive
DOS	Department of State
DOTMLPF	doctrine, organization, training, materiel, leadership and education, personnel, facilities
FM	field manual
FSF	foreign security forces
IDAD	internal defense and development
ISR	intelligence, surveillance, reconnaissance
JP	joint publication
METT-TC	mission, enemy, terrain and weather, troops and support available, time available, civil considerations (mission variables)
MTT	military transition team
NSPD	national security Presidential directive
PIR	priority intelligence requirement
S-2	intelligence staff officer
S-3	operations staff officer
S-4	logistics staff officer
SFA	security force assistance
TC	training circular
UN	United Nations
US	United States
USAID	United States Agency for International Development

SECTION II – TERMS

foreign internal defense

(joint) Participation by civilian and military agencies of a government in any of the action programs taken by another government to free and protect its society from subversion, lawlessness, and insurgency. (JP 3-05)

foreign military sales

(joint) That portion of United States security assistance authorized by the Foreign Assistance Act of 1961, as amended, and the Arms Export Control Act of 1976, as amended. This assistance differs from the Military Assistance Program and the International Military Education and Training Program in that the recipient provides reimbursement for defense articles and services transferred. (JP 1-02)

***foreign security forces**

Forces, including but not limited to military, paramilitary, police, and intelligence forces; border police, coast guard, and customs officials; and prison guards and correctional personnel, that provide security for a host nation and its relevant population or support a regional security organization's mission.

host nation

(joint) A nation that receives the forces and/or supplies of allied nations, coalition partners, and/or NATO organizations to be located on, to operate in, or to transit through its territory. (JP 3-57)

humanitarian and civic assistance

(joint) Assistance to the local populace provided by predominantly US forces in conjunction with military operations and exercises. This assistance is specifically authorized by Title 10, United States Code, section 401, and funded under separate authorities. Assistance provided under these provisions is limited to (1) medical, dental, and veterinary care provided in rural areas of a country; (2) construction of rudimentary surface transportation systems; (3) well drilling and construction of basic sanitation facilities; and (4) rudimentary construction and repair of public facilities. Assistance must fulfill unit training requirements that incidentally create humanitarian benefit to the local populace. (JP 3-57)

internal defense and development

(joint) The full range of measures taken by a nation to promote its growth and to protect itself from subversion, lawlessness, and insurgency. It focuses on building viable institutions (political, economic, social, and military) that respond to the needs of society. (JP 3-07.1)

international military education and training

(joint) Formal or informal instruction provided to foreign military students, units, and forces on a nonreimbursable (grant) basis by offices or employees of the United States, contract technicians, and contractors. Instruction may include correspondence courses; technical, educational, or informational publications; and media of all kinds. (JP 1-02)

security assistance

(joint) Group of programs authorized by the Foreign Assistance Act of 1961, as amended, and the Arms Export Control Act of 1976, as amended, or other related statutes by which the United States provides defense articles, military training, and other defense-related services, by grant, loan, credit, or cash sales in furtherance of national policies and objectives. (JP 3-57)

security force assistance

The unified action to generate, employ, and sustain local, host-nation, or regional security forces in support of legitimate authority. (FM 3-07)

security sector reform

The set of policies, plans, programs, and activities that a government undertakes to improve the way it provides safety, security, and justice. (FM 3-07)

References

Field manuals and selected joint publications are listed by new number followed by old number.

REQUIRED PUBLICATIONS

These documents must be available to intended users of this publication.

FM 1-02 (101-5-1). *Operational Terms and Graphics.* 21 September 2004.

JP 1-02. *Department of Defense Dictionary of Military and Associated Terms.* 12 April 2001.

RELATED PUBLICATIONS

These documents contain relevant supplemental information.

JOINT AND DEPARTMENT OF DEFENSE PUBLICATIONS

Most joint publications are available online: <http://www.dtic.mil/doctrine/jpcapstonepubs.htm.>

DOD 5105.38-M. *Security Assistance Management Manual.* 3 October 2003. <http://www.dsca.osd.mil/samm/>

DODD 3000.05. *Military Support for Stability, Security, Transition, and Reconstruction (SSTR) Operations.* 28 November 2005. <http://www.dtic.mil/whs/directives/corres/html/300005.htm>

JP 3-0. *Joint Operations.* 17 September 2006.

JP 3-05. *Doctrine for Joint Special Operations.* 17 December 2003.

JP 3-07.1. *Joint Tactics, Techniques, and Procedures for Foreign Internal Defense (FID).* 30 April 2004.

JP 3-07.4. *Joint Counterdrug Operations.* 13 June 2007.

JP 3-57. *Civil-Military Operations.* 8 July 2008.

ARMY PUBLICATIONS

Most Army doctrinal publications are available online: <http://www.army.mil/usapa/doctrine/Active_FM.html>. Army regulations are produced only in electronic media. Most are available online: < http://www.army.mil/usapa/epubs/index.html>.

AR 11-31. *Army International Security Cooperation Policy.* 24 October 2007.

AR 12-1. *Security Assistance, International Logistics, Training, and Technical Assistance Support Policy and Responsibilities.* 24 January 2000.

FM 3-0. *Operations.* 27 February 2008.

FM 3-05.137. *Army Special Operations Forces Foreign Internal Defense.* 30 June 2008.

FM 3-05.202 (FM 31-20-3). *Special Forces Foreign Internal Defense Operations.* 2 February 2007.

FM 3-07. *Stability Operations.* 6 October 2008.

FM 3-24. *Counterinsurgency.* 15 December 2006.

FM 3-34.400 (FM 5-104). *General Engineering.* 9 December 2008.

FM 4-0 (100-10). *Combat Service Support.* 29 August 2003.

FM 4-02 (FM 8-10). *Force Health Protection in a Global Environment.* 13 February 2003.

FM 6-0. *Mission Command: Command and Control of Army Forces*. 11 August 2003.

FM 6-22 (22-100). *Army Leadership*. 12 October 2006.

FM 7-0. *Training for Full Spectrum Operations*. 12 December 2008.

FM 7-15. *The Army Universal Task List*. 27 February 2009.

TC 31-73. *Special Forces Advisor Guide*. 2 July 2008.

OTHER PUBLICATIONS

National Defense Strategy of 2008. Washington, DC: U.S. Government Printing Office, 2008. <www.defenselink.mil/news/2008 National Defense Strategy.pdf>

National Military Strategy of the United States of America. Washington, DC: U.S. Government Printing Office, 2004. <www.defenselink.mil/news/Mar2005/d20050318nms.pdf>

National Security Strategy of 2006. Washington, DC: U.S. Government Printing Office, 2006. <www.marforres.usmc.mil/docs/nss2006.pdf>

Quadrennial Defense Review of 2006. Washington, DC: U.S. Government Printing Office, 2006. <www.defenselink.mil/qdr/report/Report20060203.pdf>

Armed Forces. Title 10, U.S. Code. <http://www.loc.gov/law/help/guide/federal/uscode.php>

Arms Export Control Act of 1976. <http://www.loc.gov/law/help/guide.php>

The United States Constitution. <http://www.house.gov/house/Constitution/Constitution.html>

Foreign Relations and Intercourse. Title 22, U.S. Code. <http://www.loc.gov/law/help/guide/federal/uscode.php>

Foreign Operations Appropriation Act of 1997. <http://www.loc.gov/law/help/guide.php>

Foreign Assistance Act of 1961. <http://www.loc.gov/law/help/guide.php>

The Geneva Conventions of 1949. <http://www.loc.gov/rr/frd/Military_Law/MLR_site-map.html>

National Security Presidential Directive 36. "United States Government Operations in Iraq." 11 May 2004.

Treaties in Force 2007. <http://www.state.gov/s/l/treaty/treaties/2007>

Charter of the United Nations. <http://www.un.org/aboutun/charter/>

War Powers Act of 1973. <http://www.thecre.com/fedlaw/legal22/warpow.htm.>

OUTSIDE RECOMMENDED READING

von Steuben, Friedrick Wilhelm. *Regulations for the Order and Discipline of the Troops of the United States*. Philadelphia: Styner and Cist, 1779.

PRESCRIBED FORMS

None

REFERENCED FORMS

DA Form 2028. Recommended Changes to Publications and Blank Forms

Index

Entries are by paragraph number.

Entries are by paragraph number.

Entries are by paragraph number.

Entries are by paragraph number.

Made in the USA
Monee, IL
05 November 2026

67808311R00164